ENTERPRISE AI FOR TREASURY

A GUIDE TO AGENTIC IMPLEMENTATION

Arjun Krishnan, CTP

FOREWORD BY BRAD LARSON

Published by Krishbooks LLC

Enterprise AI for Treasury:
A Guide to Agentic Implementation

First Edition, June 2026

Published by Krishbooks LLC

Paperback ISBN: 979-8-9961351-0-3

eBook ISBN: 979-8-9961351-1-0

Disclaimer

The views and observations expressed in this book are those of the author and do not represent the views of any organization with which the author is or has been affiliated. The proof-of-concept work described in this book was conducted independently through Valorean Technologies Inc. and is presented as exploratory research, not as production-validated performance. Readers should apply appropriate judgment when considering implementation in their own environments.

Copy Editor: Liz Wheeler

Cover and Interior Designer: Scott Moden

Printed in the United States of America

Contents

Foreword

Early in my career, a colleague once said he couldn't understand why anyone would spend hours every day producing a forecast that everyone knew would be wrong—then do the same thing again tomorrow.

I had been a corporate finance guy for about ten years when I started work at a new company and met Arjun. I had never worked with an IT person who really wanted to understand what I needed and was determined to make my job easier. We learned a lot from each other—I learned IT basics from him, and he learned a great deal about treasury needs and goals from me. That was 25 years ago. I went on to other companies, working to improve cash forecasting and other aspects of treasury. Arjun went on to build an industry-wide reputation as an IT treasury expert.

Many years ago, my CFO was drawing me into activities across every area of the company. When I asked him what his overall goal was for me, he said, "I want treasury to be involved in every decision in the company that involves money. Since every decision involves money, I want you involved in every decision in the company."

That CFO understood that treasury has a unique, unbiased view of the company. A good cash forecast is a basic building block for understanding how everything in a company comes together to make it a cohesive entity. But the ambition he described—treasury as a strategic partner in every significant decision—demands much more than better forecasting. It requires AI that can support the full breadth of treasury: liquidity management, risk, payments, hedging, and the governance frameworks that make it safe to rely on intelligent systems at all.

But numbers alone are not enough. The real opportunity AI creates for treasury is time—time freed from manual assembly and directed toward the cross-functional collaboration and strategic insight that the best CFOs have always wanted from their treasury teams. When treasury professionals spend less time gathering data and more time interpreting it, the CFO's vision becomes achievable.

I can't tell you how excited I was when Arjun told me his plans to combine his treasury technology expertise with the new tools of AI to address the challenges he has seen across the function. Cash forecasting is where many of us feel the pain most acutely, and it features prominently in this book—but Arjun's vision is larger. The data treasury needs are already sitting in various systems. He has developed a way to extract that detailed data and consolidate it into meaningful, actionable intelligence that can support AI-driven decision-making across the full treasury function.

The AI approach Arjun describes demonstrates genuine understanding of the challenges and shortcomings of current systems and manual interfaces. He knows it is an ever-improving, interactive process that constantly gets better—one that, over time, can reshape what treasury is capable of contributing to an organization.

In 2001, I joined three other corporate treasurers on a panel to discuss forecasting challenges, including a task force formed to examine available software and to answer functionality questions for users. That panel's findings led me to hope that someday we would discover the long-sought cure for the pains of cash forecasting.

It is now exactly 25 years later. Thanks to Arjun, it looks like that cure may finally be achievable—and with it, something the best CFOs have always believed treasury could become.

Brad Larson

Brad Larson is a former vice president and treasurer with more than three decades of senior leadership at Fortune 100 companies, spanning global cash management, capital markets, and enterprise risk. He has served in senior governance roles, including vice chair of the Association for Financial Professionals (AFP).

June 2026

Preface

Why I Wrote This Book

Almost three decades ago, I began my career implementing treasury systems. Since then, I've had the privilege of working with treasury teams across the globe—configuring SAP treasury modules for multinationals, helping mid-sized companies select their first treasury management system, and watching the industry evolve from paper-based processes to real-time, application programming interface (API)-driven operations. Now we stand at the next threshold: artificial intelligence-enabled treasury transformation. That evolution is what this book is about.

This book grew out of a desire to share what I've been learning about AI and its application to treasury transformation. Not as someone with all the answers, but as a practitioner working through the same questions that many treasury professionals are asking: What's real and what's hype? What's ready for production and what's still experimental? How do you implement these technologies thoughtfully, with appropriate guardrails?

The treasury profession has always been practical. We deal with real money, real deadlines, real consequences. We don't have the luxury of theoretical debates when payroll needs to be funded, or a hedging decision needs to be made. That same practicality must shape how we approach AI—not as a trend to follow, but as a capability to implement deliberately, where it genuinely adds value.

The Book is Organized in Four Parts

PART I—Chapters 1 through 4—builds the foundation: what AI actually is, how it works, where treasury fits in the current landscape, and how multi-agent systems are structured. We'll cover what's working, and what's still maturing. If you are new to AI, these chapters are worth reading in sequence. If you already have a working understanding of the technology, they serve as a reference.

PART II—Chapters 5 through 8—moves into practice: the cash forecasting and foreign exchange (FX) hedging proof-of-concept work, and the data architecture that any AI implementation depends on. We'll examine why specialized sub-agents outperform general-purpose AI, take a detailed look at intelligent cash forecasting and hedge management as examples, address the importance of data quality, and tackle the challenge of making AI explainable.

PART III—Chapters 9 and 10—addresses the control and governance layer: how to make AI systems explainable, auditable, and compliant. We'll cover how to develop practical frameworks for guardrails and explore how to build trust in autonomous systems and navigate the regulatory and compliance landscape.

PART IV—Chapters 11 and 12—is about getting started and looking forward: a practical framework for a first implementation, and my honest assessment of where treasury AI may be heading. I provide a framework using a 90-day pilot playbook as an example, and address the human side of AI adoption—how roles evolve, what skills matter, and how to bring your team along.

The book can be read end to end or by section, depending on where you are in your own exploration. Each chapter opens with a brief italic paragraph summarizing what it covers, so you can navigate directly to the sections most relevant to you. Each chapter also includes a "***Put It into Practice***" section—practical steps to start applying the ideas immediately—and a "***Chapter Summary***" section for quick reference.

Throughout, the goal is practical guidance grounded in real-world experience—the kind of insight that comes from working through actual implementation challenges, not just theorizing about possibilities.

Who This Book Is For

Whether you're a treasurer evaluating AI investments, an analyst curious about how these tools might affect your work, a CFO trying to separate signal from noise, a technology professional exploring treasury applications, or an entrepreneur building a business where cash flow is the lifeblood of growth and, for many startups, survival itself—I hope you'll find something useful here.

Treasury work, at its core, is about stewardship—managing an organization's financial resources prudently, ensuring liquidity when it's needed, and protecting against risks that could harm the enterprise. AI does not change these responsibilities. What it changes is the scale, speed, and intelligence with which they can be fulfilled. Understanding that distinction—what AI transforms and what it doesn't—is one of the central threads of this book.

My hope is that this book helps you engage with AI-driven treasury transformation thoughtfully, practically, and with appropriate skepticism. The technology is genuinely powerful. Implementing it well requires wisdom that no algorithm can provide.

That wisdom comes from practitioners who understand both the potential and the pitfalls of applying AI in a domain where the stakes are real—treasury professionals who bring judgment, experience, and context to decisions that matter.

In other words, it comes from people like you.

That's the spirit in which this book is offered. Let's begin.

Arjun Krishnan, CTP

June 2026

For my grandchildren,
who make my every
day worth living.

> “We can only see a short distance ahead, but we can see plenty there that needs to be done.”
>
> **Alan Turing**
>
> Computing Machinery and Intelligence (1950)

PART I

"It is a capital mistake to theorize before one has data."

Sir Arthur Conan Doyle
The Adventures of
Sherlock Holmes (1892)

CHAPTER 1

The Autonomy Spectrum

Where Treasury Technology Is—and Where It's Headed

This chapter introduces a four-stage framework, based on the concepts of rules, analytics, assistance, and autonomy. The Autonomy Spectrum maps where any treasury operation sits in its technology journey. It gives you a practical tool for assessing your own organization's current position honestly, and for thinking about what a realistic next step looks like. Most treasury teams are further along than they think on at least some dimensions.

Almost three decades of working at the intersection of treasury and technology have taught me that the most important questions are rarely technical. Every treasury transformation, across every organization and every era of technology, has come down to the same underlying challenge: understanding where you are before deciding where to go. The teams that navigate change well are not necessarily the ones with the biggest budgets or the most sophisticated systems. They are the ones that assess their current position honestly, set realistic targets, and move deliberately—closing the gap one step at a time.

That's what this chapter is about. Before diving into AI architectures and machine learning models, it helps to have a shared framework for thinking about automation, intelligence, and autonomy—a map of the territory that makes it easier to assess where your organization sits today, and where it might reasonably go.

The Reality of Treasury Operations Today

Walk into almost any corporate treasury center and you'll find a familiar pattern: sophisticated technology surrounded by manual workarounds.

Treasury management systems (TMSs) are software that help companies manage cash, payments, and financial risk; your organization's TMS consolidates bank balances automatically. But every morning, an analyst downloads the data into a spreadsheet to build the daily cash position report, because the standard reports don't quite match what the CFO wants to see.

The payment system processes thousands of transactions through straight-through processing. But every afternoon, someone reviews a manually generated exception report, eyeballing each flagged item against an informal checklist that exists only in their head.

Bank statements arrive via SWIFT (the Society for Worldwide Interbank Financial Telecommunication—the global network banks use to exchange financial messages securely), posting automatically to the general ledger. But month-end reconciliation still requires treasury staff to work late, matching transactions between systems that should talk to each other, but often don't.

These are just a few examples. Consider the broader landscape:

Cash forecasting often relies on Excel workbooks maintained by individual analysts. Subsidiaries email their forecasts (if they remember), which get manually consolidated. Historical accuracy is tracked inconsistently, if at all.

Intercompany netting may be calculated in the enterprise risk planning (ERP) software, which manages core business processes, but the settlement process involves manual journal entries, email confirmations, and reconciliation spreadsheets that circulate between shared services centers.

FX exposure management means pulling position reports from multiple systems, consolidating them in Excel, comparing the results to policy thresholds, manually keying hedge orders into a trading platform, and then recording and reconciling trades back into the system of prime record.

Bank fee analysis happens quarterly (if it happens at all), requiring someone to download statements, parse cryptic codes, and build a comparison spreadsheet from scratch each time.

The technology does a lot. The humans fill in the gaps.

This isn't a criticism—it's simply the reality of complex organizations with legacy systems, ERP systems that never quite fulfill their true potential, diverse requirements, and finite IT resources. Treasury teams may be small, but are remarkably good at making things work.

But there's a cost. That cost shows up in analyst hours spent on data manipulation rather than analysis. It shows up in errors that slip through manual processes. It shows up in decisions made with yesterday's information because today's isn't quite ready yet. And it shows up in the talented professionals who spend their days on tasks that don't use their real expertise.

What Becomes Possible with Agentic AI

Now consider how these same challenges might be addressed with a different approach.

What if the daily cash position report generated itself? Not just pulling numbers from the TMS, but understanding what the CFO wants to see, adapting the format based on what's most relevant today, and highlighting the items that need attention.

What if payment exceptions weren't just flagged, but triaged? A system that learns from thousands of prior decisions could assign risk scores, suggest appropriate actions, and route only the genuinely unusual cases to human reviewers.

What if intercompany netting weren't just calculated, but orchestrated? From identifying netting opportunities across currencies and time zones, to generating settlement instructions, to reconciling the results—all coordinated automatically with appropriate checkpoints for approval.

What if FX exposure management happened continuously rather than periodically? Positions could be monitored in real time, policy limits tracked automatically, hedging recommendations generated when thresholds approach, and execution initiated when pre-approved parameters are met.

This is what becomes possible when AI moves beyond analysis and into action. Not replacing human judgment, but augmenting it—and handling the routine so that human judgment can focus on what really matters.

The technology for this exists today. The question is how to implement it thoughtfully—with appropriate guardrails, realistic expectations, and a clear understanding of where human oversight remains essential.

The Autonomy Spectrum

It's useful to think about treasury technology along a spectrum with four distinct stages: rules, analytics, assistance, and autonomy. This framework can help assess where different processes stand today and where they might reasonably go.

Let's walk through each stage, because understanding where your organization sits on this spectrum is the starting point for any AI initiative.

Stage 1: Rules

At the rules stage, technology follows explicit instructions written by humans. If the balance exceeds a threshold, sweep the excess. If the payment is under a certain amount and to an approved vendor, auto-approve. If the invoice matches the purchase order (PO) within tolerance, process it.

This is the world of robotic process automation (RPA)—software that mimics human actions to perform repetitive tasks—of workflow engines, and of business rules configured in your ERP system or TMS. The system does exactly what you tell it, every time, without deviation or judgment.

The strength of rules: predictability, auditability, and reliability. When a rule-based system processes a million payments and none of them fail, you can trust it with the next million.

The limitation of rules: rigidity. Someone must anticipate every scenario and write a rule for it. When something unexpected happens—and in treasury, unexpected things happen constantly—the system stops and waits for a human.

Roughly half of treasury operations globally are still predominantly in the rules stage. This reflects the genuine value that well-implemented automation delivers. Rules-based automation has eliminated manual keying errors, accelerated processing times, and freed treasury professionals from the most repetitive tasks.

But rules can only take you so far.

Stage 2: Analytics

At the analytics stage, technology provides visibility into what has happened. Dashboards consolidate data from multiple sources. Reports show trends over time. Visualizations make patterns apparent that would be invisible in raw numbers.

This is the world of business intelligence (BI)—tools and practices for collecting, analyzing, and presenting business data—of data warehouses, of the interactive dashboards that help treasury teams understand their position at a glance.

The strength of analytics: transparency, understanding, and insight. When you can see your global cash position in real time, color-coded by currency and entity, you make better decisions than when you're working from yesterday's spreadsheet.

The limitation of analytics: passivity. The analytics dashboards tell you what happened. They don't tell you what to do about it. The human still performs all the real, complex analysis, makes the important decisions, and exercises informed judgment. The dashboard is a window, not an advisor.

Another 30% or so of treasury operations are primarily in the analytics stage. They've invested in visibility. They've connected their data sources. They have sophisticated, often customized reporting. But when you watch their treasury professionals work, you see people evaluating the information on the screens, synthesizing information, and making decisions in their heads.

The technology informs. The human decides.

Stage 3: Assistance

This is the emerging world of AI copilots, recommendation engines, and intelligent decision-support systems. Their role is not yet to act independently, but to analyze the available data, interpret patterns, and recommend a course of action for human validation. In treasury, this may mean proposing an investment allocation, identifying hedge candidates, or flagging invoices likely to be paid late. As these capabilities become more deeply embedded into workflows, some may form the foundation for agentic systems that can coordinate tasks and operate within defined controls—a progression covered in "Stage 4: Autonomy."

The strength of assistance: leverage. The system does much of the analytical work that previously required senior expertise. A junior analyst can now consider options that previously only the treasurer would evaluate, because the AI has done the heavy lifting.

The limitation of assistance: bottlenecks. A human is still required to make every decision. They still click "approve" or "reject" on every recommendation. As the system gets smarter and makes more recommendations, the human becomes the constraint.

Perhaps 15% of treasury operations have reached the assistance stage. They're using AI to suggest actions. They're getting recommendations from their systems. But they're not yet trusting those systems to act.

Stage 4: Autonomy

At the autonomy stage, technology observes, decides, and acts within defined boundaries. The system doesn't just recommend a cash positioning strategy—it executes it. The fraud detection system doesn't just flag suspicious transactions—it holds them automatically. The forecasting engine doesn't just predict a shortfall—it initiates the draw on the revolver.

This is the frontier. This is where AI becomes agentic—pursuing goals rather than following instructions, taking action rather than making suggestions.

The strength of autonomy: scale and speed. Autonomous systems can make thousands of decisions per day that no human team could review. They can act in milliseconds when markets move. They can process complexity that would overwhelm any analyst.

The limitation of autonomy: trust. When you let a system act on your behalf, you need confidence that it will do the right thing, even in scenarios you haven't imagined. You need guardrails that contain the damage when—not if—it makes a mistake (more on this in Part III).

Perhaps 5% of treasury operations are employing autonomous AI systems in a meaningful way, mostly in narrow domains like fraud screening or outlier detection. The technology is capable of more, but building the organizational trust to deploy it takes time.

Where Are You on the Spectrum?

Take a moment to think about your own treasury operation. For each major process, where would you place your organization? Use **TABLE 1.1**.

TABLE 1.1: Where does your organization sit on the maturity spectrum?

Process	Rules	Analytics	Assistance	Autonomy
Cash positioning				
Cash forecasting				
Payment processing				
Fraud detection				
FX exposure management				
Bank relationship management				
Liquidity planning				

Most treasury professionals find that they're in different stages for different processes. Cash positioning might be well-automated (rules), while cash forecasting is still spreadsheet-based (not even fully in analytics). Fraud detection might have some AI screening (assistance), while FX management is entirely manual.

This is normal. The spectrum isn't something you cross all at once. It's a process-by-process journey.

Most treasurers are in the first two stages. That's not "behind"—it reflects where the industry is. The goal isn't to rush to autonomy; it's to move deliberately, building trust and managing risk at each step.

The Critical Threshold

The jump from the assistance to autonomy stages represents a fundamental shift worth understanding clearly.

In Stages 1 through 3, the human remains in control. Rules are executed, but humans wrote them. Analytics software informs, but humans interpret the outputs. AI assistants make recommendations, but humans decide.

In Stage 4, the machine decides. Within boundaries, yes. Subject to guardrails, certainly. But the locus of decision-making shifts from human to artificial intelligence.

This shift triggers every concern that people have about AI: machines making decisions with real consequences, job evolution, changing skill requirements. These concerns aren't irrational. They're reasonable responses to a genuine change in the relationship between humans and technology.

What I've observed in practice is that these concerns are manageable when addressed honestly, and the benefits are substantial when implementation is thoughtful.

The key insight is that autonomy isn't binary. It's not "the AI controls everything" or "humans control everything." It's a calibrated trust relationship, where autonomy is granted in proportion to demonstrated reliability, with guardrails proportional to risk, and with humans very much "in the loop."

We'll explore this in depth in later chapters. For now, the central idea is this: autonomy is earned, not granted. Systems graduate from the assistance to autonomy stages through established track records and earned trust.

Why It's Crucial to Consider the Autonomy Spectrum Now

Some parts of the Autonomy Spectrum have been around for a long time. Recent developments in AI, however, are making the transition from the assistance stage to the autonomy stage feasible for the first time.

First, the AI itself has gotten dramatically better. The large language models (LLMs)—AI systems trained on vast amounts of text that can understand and generate human language—that power tools like ChatGPT and Claude represent a step change in machine reasoning capability. For the first time, we have AI that can understand context, handle ambiguity, and make judgment calls that previously required human intelligence. This isn't an incremental improvement; it's a qualitative transformation.

Second, the integration infrastructure has matured. APIs—standardized ways for different software systems to communicate—are everywhere. Cloud platforms enable real-time data sharing. The technical barriers to connecting AI systems with treasury systems have never been lower. An AI agent that can reason about your cash position can now actually do something about it, because the connections exist to make it happen.

Third, the data is finally available. Twenty years of treasury technology investment have created unprecedented data assets. Transaction histories, cash position records, and forecast accuracy tracking provide raw material for training machine learning (ML) systems, which AI is based on. These systems learn patterns from data rather than following explicit rules, and the data they need to learn from has been collecting for two decades, waiting to be used.

The technology has crossed a threshold. The question is no longer "Can AI handle treasury decisions?" The question is "Which decisions, under what conditions, and with what safeguards?"

Navigating the Noise

Anyone paying attention to financial technology over the past two years has noticed a surge of excitement around AI. Wall Street analysts are projecting massive market growth. Industry publications run weekly features on AI transformation. Conference agendas are packed with sessions on artificial intelligence.

Some of this enthusiasm is well-founded. The underlying technology has genuinely improved, and the potential applications in treasury are real.

But some of the excitement has outpaced reality. Projections sometimes conflate what's possible in a research lab with what's deployable in an enterprise treasury environment. Timelines can be optimistic. The gap between a compelling demo and a production-ready system is often larger than it appears.

Navigating that noise honestly requires acknowledging something upfront: the AI landscape is new enough, and evolving fast enough, that no one—including the author—has all the answers. What these pages offer is an honest account of what careful exploration has revealed: which approaches show genuine promise, where the gap between demo and deployment remains wide, and how treasury professionals can begin thinking through these challenges for their own organizations. What this book will not do is promise that AI will solve every problem or that its implementation is straightforward. The real transformation happens when these ideas are applied to specific systems, specific data, and specific constraints—and that requires collaboration between practitioners, not proclamations from any single authority.

PUT IT INTO PRACTICE

In this section, I'll show you how to map your treasury processes against the Autonomy Spectrum.

- Refer to the table shown earlier; fill it in. For each major process in your treasury operation, assess where you are today. Be honest. Most processes will be in the first two stages. That's fine—it's where most of the industry is.
- Then, for each process, ask yourself: where should it be in two years? Where is the biggest gap between my organization's current state and ideal state? Where would moving one stage along the spectrum deliver the most value?
- This exercise takes perhaps an hour. But it will frame everything else you read in this book. You'll be evaluating concepts against your specific situation, not abstract possibilities.
- And that's how AI is most effectively approached: not as a general trend to follow, but as a specific capability to apply where it matters most for your organization.

CHAPTER SUMMARY

- Treasury technology evolves through four stages: rules (follows instructions) → analytics (provides visibility) → assistance (makes recommendations) → autonomy (decides and acts).
- Most treasury operations are in Stages 1–2; about 15% have reached Stage 3; only ~5% are in Stage 4—and that's perfectly normal.
- The jump from the assistance to autonomy stages is the critical threshold: where the locus of decision-making shifts from human to AI.
- AI autonomy should be earned through demonstrated reliability, not granted all at once, and always with humans in the loop.
- Three converging factors make this transition newly possible: better AI, mature integration infrastructure, and available data.
- Navigate the market excitement with humility—the AI space is still evolving, and honest exploration matters more than definitive prescriptions.

In the next chapter, we'll go deeper on what AI is—understanding the capabilities and limitations of the technology that's reshaping our industry.

CHAPTER 2

Understanding AI

What It Is, What It Does, and Why It Matters for Treasury

AI is one of the most discussed terms in enterprise technology, and one of the most variably defined. This chapter explains what AI is, how the main types work, and where each has found practical application in a treasury context. It is written as a snapshot of where the technology stands in mid-2026—a foundation for understanding everything that follows.

If you've made it through the first chapter, you have a framework for thinking about where treasury technology is headed. But we've been talking about AI without really defining it. This chapter addresses that gap.

The term "artificial intelligence" is used to describe an enormous range of capabilities—from the basic automation rules already embedded in most treasury management systems, all the way to artificial general intelligence (AGI), the idea of a system that can perform any intellectual task a human can (and more), which remains a distant and still-debated prospect. In between sits a wide and rapidly evolving spectrum of practical tools, and the terminology that describes them continues to evolve just as quickly.

The goal of this chapter is to offer a working understanding of AI that is accurate enough to be useful and grounded enough to apply in practice—when evaluating a forecasting system, assessing a new treasury application, or thinking through which treasury processes might benefit from AI-based approaches. A treasurer who understands these concepts clearly is better placed to engage with the technology on their own terms.

Let's start with what AI actually is.

A Working Definition of AI

Here's a definition that I find useful: AI is software that can make decisions in situations it wasn't explicitly programmed to handle.

Traditional software follows rules: if this condition, then that action. The programmer anticipates scenarios and writes instructions for each one. The software does exactly what it's told, nothing more.

AI software learns patterns from data, then applies those patterns to new situations. It recognizes what it hasn't seen before by finding similarities to what it has seen. It makes judgment calls, not just rule applications.

This distinction matters for treasury. Let's consider a concrete example.

When you configure a payment approval workflow—"amounts over $50,000 require CFO sign-off"—that's traditional software following explicit rules. Someone anticipated this scenario and wrote an instruction for it.

But when a system flags a payment as potentially fraudulent because it resembles patterns learned from thousands of historical fraud cases—even though this specific payment doesn't match any predefined rule—that's AI. No one wrote a rule for this exact scenario. The system identified it from examples.

The power of AI comes from this ability to generalize. You don't have to anticipate every fraud scenario and write a rule for it. You show the system examples of fraud and non-fraud, and it learns to tell the difference—including in situations that may not have been encountered before.

The limitations also come from this capacity to generalize. Because the system learned patterns rather than following rules, you can't always predict exactly what it will do. It might flag legitimate payments. It might miss novel fraud schemes. The judgment calls are real, and like all judgment calls, they're sometimes wrong.

Three Types of AI Relevant to Treasury

In treasury, the term "AI" can refer to any of three quite different categories of technology. Understanding the differences matters when assessing capabilities, evaluating whether a particular approach fits a specific use case, and forming a view on what is genuinely ready for implementation.

Machine Learning: Pattern Recognition at Scale

Machine learning is one of the workhorses of practical AI. It enables systems to improve their performance on a task by learning patterns from data, rather than relying solely on manually programmed rules. In enterprise applications, ML is often combined with business rules, controls, and human judgment to produce recommendations that are both data-driven and operationally usable.

Here's how it works in a treasury context: suppose you want to predict which customer invoices will be paid late. You gather historical data—thousands of invoices with their characteristics (amount, customer, payment terms, time of year, industry, and so on) and their outcomes (paid on time or late). You provide this to an ML algorithm, which finds patterns.

Perhaps invoices from certain customers tend to be late. Perhaps large invoices in December are slower to pay. Perhaps invoices without PO numbers have higher delinquency rates. The algorithm discovers these patterns automatically.

The output is a model—essentially a mathematical formula, or algorithm in AI terms—that takes invoice characteristics as input and produces a lateness probability as output. When a new invoice arrives, you run it through the model, and might get a response like this: "This invoice has a 73% chance of late payment based on historical patterns."

Common ML applications in treasury include cash flow forecasting, fraud detection, payment behavior prediction, and transaction categorization and matching.

ML requires three things: data to learn from, computational power to process it, and expertise to build and tune the models. The data requirement is often the binding constraint—ML is only as good as the examples it learns from. If your historical data doesn't include examples of a particular fraud or error type, the system won't learn to detect it.

Natural Language Processing: Understanding Text

Natural language processing (NLP) is AI that works with human language—reading, writing, and understanding text.

For treasury, this matters in several practical areas. When a bank confirmation or bank statement arrives as a PDF file, NLP can automate the extraction of confirmation numbers, amounts, dates, and counterparty details rather than requiring manual data entry into the TMS. For credit agreements, International Swaps and Derivatives Association (ISDA) masters, and bank fee schedules, NLP can help identify key terms, flag unusual provisions, and extract structured data from dense documents. Treasury operations that receive large volumes of emails—payment instructions, bank notifications, trade confirmations, SWIFT messages, vendor queries—can use NLP to categorize and route these automatically. And conversational interfaces that allow a treasurer to ask something like "What's our EUR exposure by entity?" and receive a direct answer are becoming practical with modern NLP.

The capability of NLP systems has advanced significantly in recent years. Modern systems, built on transformer architectures—the technology behind tools like ChatGPT—can understand context, handle ambiguity, and process language in ways that were not practical just a few years ago.

Large Language Models: General-Purpose Reasoning

LLMs are the newest and most discussed category. These are AI systems trained on vast amounts of text that can understand and generate human language with considerable sophistication.

What makes LLMs different from traditional NLP systems is their generality. A traditional NLP system might be trained specifically to extract payment terms from contracts. An LLM can do that, but it can also answer questions about the contract, summarize it, translate it, compare it to other contracts, and have a conversation about its implications—all without being specifically trained for any of these tasks.

Emerging LLM applications in treasury include natural language interfaces to treasury systems, automated report generation and narrative explanation, analysis and interpretation of complex data, and decision support through reasoning and explanation. For instance, rather than just showing a cash forecast chart, an LLM-powered system might explain: "Cash is projected to be tight in week three because the quarterly tax payment coincides with a large vendor payment. Consider whether delaying the vendor payment or drawing on the revolver is appropriate."

The limitation of LLMs is that they're probabilistic—they generate responses based on statistical patterns in their training data. They can be confidently wrong. They can produce plausible-sounding but incorrect outputs, a phenomenon sometimes called "hallucination." They need guardrails and verification, particularly in treasury, where accuracy is essential.

What's Production-Ready, What's Maturing, and What's Experimental

Not all AI is equally ready for treasury application, and it is worth being clear about where different capabilities sit on that maturity spectrum.

Production-Ready Today

These applications have established track records in enterprise environments:

Straight-through processing and rule-based automation. The use of robotic process automation (RPA) —software that mimics human actions to perform repetitive tasks—and rule-based workflow automation in treasury is well-established. Payment straight-through processing, automated bank reconciliation, and rule-based exception routing have been in production for years. These are foundational applications that most well-configured treasury operations already use in some form.

Transaction categorization and matching. ML systems that classify transactions, match payments to invoices, and identify exceptions have been in production for over a decade. The technology is mature, the accuracy is high, and the failure modes are well understood. A common example: automatically matching incoming payments to open invoices based on learned patterns, even when reference numbers are missing or formatted inconsistently.

Anomaly detection for payment fraud. Systems that flag unusual transactions based on learned patterns are widely deployed in payment processing. They are not perfect—false positives are common—but they are reliable enough for production use with human review. For example: flagging a wire transfer to a new beneficiary when the pattern differs from that vendor's historical behavior.

AI-assisted daily operations. AI tools that support routine treasury tasks are increasingly available—automated cash position reporting that aggregates bank feeds and TMS data, intelligent exception queues and alerts that prioritize items requiring attention, and automated reconciliation that reduces the manual effort of matching positions across systems.

Maturing Rapidly

These applications work well in controlled settings but require careful implementation:

AI-assisted cash flow forecasting. ML-based forecasting approaches show meaningful accuracy improvements in controlled implementations, but cash forecasting in most organizations still relies heavily on Excel, manual subsidiary submissions, or heavily customized TMS configurations. AI-based forecasting is a genuinely promising area—and one this book explores in depth in Chapters 5 and 6—but describing it as fully production-ready would overstate where most implementations are today.

Document extraction. NLP systems can extract data from both bank statements in standard formats (MT940, camt.053) and bank confirmations for treasury deals with good accuracy, but documents in non-standard or proprietary formats may still require human review. The technology is improving quickly.

Natural language queries. Asking questions in plain language and receiving useful answers is becoming practical, but systems still struggle with ambiguity. "Show me our largest FX exposures" might work reliably; "What should I be most concerned about this week?" is harder to answer consistently.

Predictive analytics beyond forecasting. Predicting customer payment behavior, counterparty risk, or market movements works in some contexts but requires significant data quality and implementation expertise.

Experimental— Proceed with Appropriate Care

These applications show promise but are not yet reliable enough for unsupervised production use:

Autonomous decision-making. Systems that make and execute decisions without human approval require robust guardrails outside narrow, well-defined domains. In treasury, this includes decisions such as automatically hedging exposures, repositioning cash, or adjusting investment allocations — areas where the consequences of an unchecked error are material.

Complex document generation. LLMs can draft documents—board memos, covenant compliance reports, policy updates—but the output requires careful review. Errors can be subtle and consequential.

Strategic recommendations. AI can support analysis effectively, but relying on it for strategic decisions, such as those about capital structure, financing structure, or long-term planning, without expert validation is premature. The risk is that it optimizes for the wrong objective or misses the organizational context that an experienced treasurer would naturally apply.

Complex negotiation support. Using AI to support negotiations around credit agreements, banking arrangements, or counterparty contracts is emerging but not mature. The nuance required in these interactions exceeds current AI capabilities.

How AI Systems Learn

Understanding the learning process helps when evaluating AI systems and their limitations. There are three main approaches, each suited to different types of treasury problems.

Supervised Learning: Learning from Examples

You provide the system with labeled examples: "This transaction is fraud; this one isn't." "This customer paid late; this one paid on time." The system learns to distinguish between categories based on the examples you give it.

An example in the context of treasury would be training a model to predict late payments by providing it with historical invoices labeled with whether they were paid on time or late (based on due date versus paid date). The model learns to recognize the characteristics that correlate with late payment.

The key consideration is this: how representative are your examples? If your historical data doesn't include examples of a particular pattern, the system won't learn to recognize it. If your examples are concentrated in certain customer segments, predictions will reflect that.

Unsupervised Learning: Finding Structure

Unsupervised learning involves providing the system with data without labels and asking it to find patterns on its own. Rather than learning from labeled examples of fraud, the system learns what "normal" payment patterns look like and flags transactions that deviate significantly from that baseline.

This is a different approach to fraud detection from supervised learning. Supervised learning asks: does this transaction resemble the fraud cases I've seen before? Unsupervised learning asks: does this transaction look unusual compared to the patterns I've learned from normal activity? Both approaches have their place. Supervised learning is effective when you have good labeled examples. Unsupervised learning is valuable for detecting novel patterns that have no prior examples—including new fraud techniques.

An example in the treasury context would be training a model on historical payment patterns without labeling anything as fraud. The system learns what typical payments look like and flags transactions that deviate significantly—a new beneficiary, an unusual amount, an unexpected time of day.

The key consideration is this: does the system's definition of "normal" match yours? It might flag patterns you don't consider concerning while missing schemes that mimic normal behavior.

Reinforcement Learning: Learning from Feedback

With reinforcement learning, the system takes actions, receives feedback (rewards or penalties), and adjusts its behavior to maximize rewards over time.

An example in the context of treasury would be a cash positioning system that makes investment decisions and receives feedback based on returns achieved. Over time, it learns strategies that maximize return within defined risk constraints.

The key consideration is this: are you measuring the right thing? If the system is rewarded for maximizing yield, it might take risks you didn't intend. Getting the feedback function right requires careful thought.

Key Limitations to Understand

Every technology has limitations. AI's limitations are worth understanding clearly because they can be less visible than those of traditional software.

Data quality shapes AI quality. AI learns from data. If your data is incomplete, biased, or inconsistent, the AI will learn those characteristics. Unlike traditional software, where data errors tend to produce obviously wrong outputs, AI can produce plausible-seeming outputs that are subtly wrong. A cash forecasting model trained on historical data where certain subsidiaries consistently submitted late—with actuals recorded instead of forecasts—will learn that those subsidiaries are highly predictable, when really it has learned the reconciliation pattern rather than genuine business behavior.

AI correlates; it doesn't understand. When an LLM answers a question about treasury policy, it is not reasoning from first principles. It is finding patterns in text that resemble the question and generating responses that tend to follow such patterns statistically. This works well much of the time. But the system does not understand treasury the way an experienced professional does. It can miss context, make confident errors, and encounter situations where the right response requires judgment that the model has not developed. This is precisely why keeping humans in the loop is not just a governance preference in treasury AI—it is a practical necessity. The treasury professional who reviews an AI recommendation brings organizational knowledge, relationship context, and judgment that no current model can fully replicate.

AI struggles with genuinely novel situations. AI systems perform well on data that resembles their training data. They struggle with situations that are genuinely novel. An ML model might handle routine payment fraud detection well but fail to recognize a sophisticated business email compromise scheme it has never encountered—exactly the case where catching it matters most. The cases where AI is most likely to struggle are often the highest-stakes ones.

AI behavior can vary with inputs. Traditional software behaves the same way every time you run it. AI systems can behave differently based on subtle variations in input. A small change in how you phrase a question might produce a significantly different answer. This stochastic behavior requires different approaches to testing and validation than traditional software.

Explainability remains challenging. Complex AI models are often difficult to explain. You can see what the system decides, but not always why. For treasury, where audit trails and explainability matter for both internal governance and regulatory purposes, this is a real consideration. There are techniques for making AI more interpretable, but they involve trade-offs. We explore this in Chapter 9.

Economic Considerations

Beyond technical capability, AI implementation has economic dimensions that affect feasibility.

Build vs. buy. Building custom AI requires expertise (data scientists, ML engineers), data infrastructure, and ongoing maintenance. Most treasury organizations do not have these resources in-house, and acquiring them is expensive. Buying AI means evaluating vendor offerings, integrating with existing systems, and accepting choices about model architecture and training data that the vendor has made. For most organizations, a hybrid approach makes sense: buy platforms with embedded AI for standard use cases, build custom systems—or partner with specialists—for differentiated capabilities that provide specific competitive advantages.

Data as a strategic asset. AI learns from data. Organizations with more data, cleaner data, and longer history can build better models. This creates a compounding effect: better predictions lead to better decisions, which generate more data, which enables even better predictions. Capturing and organizing treasury data systematically is not just a reporting investment—it is a foundation for future AI capability.

Total cost of ownership. The cost of AI extends well beyond the software license. Data preparation and cleansing, integration with existing systems, training and change management, ongoing model monitoring and retraining, and any expert oversight that AI reduces but does not eliminate—all of these are real costs that belong in any honest evaluation of an AI investment.

PUT IT INTO PRACTICE

Take an hour this week to assess your treasury operation's current relationship with each of the three AI types described in this chapter.

- Where is ML already operating in your environment? Payment matching, fraud screening, and forecasting are the most common starting points.
- Where is your most significant data asset? Which treasury data set is cleanest, most consistent, and most accessible? That is your most practical starting point for any AI initiative. The data you have determines what is achievable.
- Where are your team's highest-friction manual processes? The treasury tasks that consume the most time with the least judgment required are often the strongest early candidates for AI-assisted automation. Consider mapping three to five of those processes against the Autonomy Spectrum from Chapter 1.
- What would "good" look like? For each process you identify, describe what the outcome would look like if AI were supporting it well. A clear vision of the goal is as important as an understanding of the technology.

CHAPTER SUMMARY

- AI is software that makes decisions in situations it wasn't explicitly programmed for—it learns patterns from data rather than following rules.
- Three types of AI matter for treasury: machine learning (pattern recognition from examples), natural language processing (understanding and generating text), and large language models (general-purpose reasoning).
- AI maturity varies significantly by use case: straight-through processing, transaction matching, and fraud detection are production-ready; AI-assisted cash forecasting and document extraction are maturing; autonomous decision-making remains experimental.
- AI correlates rather than understands—it finds patterns rather than reasoning from first principles; keeping humans in the loop is a practical necessity in treasury, not just a governance preference.
- Key limitations include data quality dependence, difficulty with novel situations, variable behavior with similar inputs, and explainability challenges.
- Economics matter: the build vs. buy decision, data as a long-term strategic asset, and total cost of ownership—all belong in any honest AI investment evaluation.

In the next chapter, we'll survey the current landscape of AI applications in treasury—what's working, what's emerging, and what's still over the horizon. We'll move from understanding AI in the abstract to seeing how it's being applied in practice.

CHAPTER 3

The Treasury AI Landscape

What's Working, What's Emerging, and What's Still Over the Horizon

This chapter surveys AI across the major treasury use cases—cash forecasting, fraud detection, payments, FX risk, reconciliation, and more—and gives each an honest maturity assessment: production-ready, maturing, or emerging. The goal is a clear and realistic picture of where value is available now, where it is coming, and what is still developing. The assessments that follow reflect the state of AI in treasury as of mid-2026. This is a landscape that is moving quickly—the categories are more durable than the specific examples within them, and readers coming to this chapter in future years should treat the maturity ratings as a baseline rather than a current verdict.

In Chapter 2, we explored what AI is and how it works. Now let's turn to a more practical question: where is AI actually being applied in treasury today, and what results is it delivering?

This chapter surveys the landscape across the major treasury use cases—cash forecasting, fraud detection, payment processing, document extraction, FX hedge management, bank fee analysis, reconciliation, and working capital optimization. For each, the goal is an honest assessment of where it sits on a three-stage maturity spectrum: production-ready, maturing, or experimental.

In Chapter 2, we introduced this three-stage framework. It is worth keeping it in mind as we move through the use cases here, because not every application is equally ready for deployment. The distinction matters when setting expectations, prioritizing where to start, and deciding how much organizational investment a particular initiative warrants.

Some of what follows will be familiar territory. The purpose is not to catalogue every possible AI application, but to give treasury professionals a grounded, honest picture of where the technology actually is—not where the most optimistic projections place it.

Cash Flow Forecasting

Cash forecasting is the treasury function where AI-based approaches have generated the most active exploration and the most promising early results. The reasons are straightforward: forecasting is pattern recognition at its core, historical data is typically available in some form, and even modest accuracy improvements translate directly into better funding, investment, and liquidity decisions.

What AI Brings to Forecasting

Traditional forecasting approaches—spreadsheet models, simple averages, manual adjustments based on experience—struggle with complexity. When cash flows are influenced by dozens of variables (seasonality, payment terms, customer behavior, economic conditions, billing cycles), tracking all the interactions reliably is genuinely difficult. Machine learning models can identify patterns across many variables simultaneously, detect seasonality at multiple frequencies, and adapt as patterns change over time.

A practical example: consider a company with significant seasonality, varying payment terms across customer segments, and periodic large transactions such as tax payments, dividend distributions, and debt service. A traditional model might handle seasonality reasonably well but struggle when multiple factors interact—say, when a seasonal peak coincides with a large scheduled payment. An ML model trained on historical data learns these interactions automatically.

Where Cash Forecasting Sits on the Maturity Spectrum

Cash forecasting AI is—in my assessment—best described as "maturing rapidly" rather than fully production-ready. The potential is genuine and the proof-of-concept work, described in Chapters 5 and 6, produced encouraging results. But in practice, most organizations still rely heavily on spreadsheets, manual subsidiary submissions, or heavily customized TMS configurations for their forecasting. The gap between a controlled implementation on clean data and a reliably operating system across the full complexity of a real multinational treasury is significant—and worth being honest about here.

Chapters 5 and 6 address this in depth: the ensemble approach, what proof-of-concept exploration revealed at each time horizon, and what genuinely seems within reach versus what remains ahead. This section is deliberately brief in recognition of that fuller treatment.

What to Watch For

Forecasting AI is only as good as the data it learns from. If historical data has quality issues—inconsistent categorization, missing periods, accounting adjustments that distort patterns—the model will learn those problems. Data preparation typically consumes most of the implementation effort. Models also require ongoing monitoring: customer behavior shifts, business models evolve, and models that performed well historically can degrade if not periodically retrained.

Fraud Detection and Payment Screening

Fraud detection is an area where AI has demonstrated consistent, tangible value. The combination of high stakes, pattern-based threats, and the need for real-time decisions makes this a natural fit for machine learning.

How AI Detects Fraud

Traditional fraud detection relies on rules: flag payments over a certain amount, to certain countries, or to new beneficiaries. These rules catch known patterns but struggle with novel threats. ML-based fraud detection takes a different approach. Rather than encoding specific rules, the system learns what "normal" looks like for your organization—

typical payment amounts, usual beneficiaries, expected timing patterns, standard approval workflows. Transactions that deviate significantly from normal are flagged for review.

A practical example: a business email compromise attack might result in a payment that passes all traditional rule checks—it's under the threshold, to an existing counterparty country, with proper approvals. But the ML system notices subtle anomalies: the payment timing is unusual for this counterparty, the amount is slightly different from the typical pattern, the beneficiary account was recently changed. Any one factor might not trigger a rule; together, they raise the risk score enough to warrant human review.

Where Fraud Detection Sits on the Maturity Spectrum

Payment fraud detection is—genuinely—"production-ready" and widely deployed. Most treasury management systems and banking platforms incorporate some form of ML-based screening. The technology works well but is not perfect. False positives are common. The practical challenge is tuning the system to catch genuine threats without creating so many false alerts that the treasury team becomes overwhelmed.

What to Watch For

Fraud detection systems need continuous learning. Attack patterns evolve constantly, and a system trained on last year's fraud attempts may not recognize this year's techniques. AI-based fraud detection is also a complement to, not a replacement for, strong controls. Segregation of duties, approval workflows, and verification procedures remain essential. AI adds a layer of pattern-based detection but does not eliminate the need for fundamental controls.

Payment Processing and Optimization

Beyond fraud detection, AI is increasingly applied to payment operations more broadly—routing optimization, timing decisions, exception handling, and reconciliation.

Payment Routing and Timing

For organizations with multiple payment channels and banking relationships, AI can optimize which payments go through which channels to minimize costs and maximize efficiency. The system learns cost structures, processing times, and reliability patterns across channels and makes routing recommendations. Similarly, AI can support payment timing decisions—when same-day versus next-day processing makes sense, when to batch versus send individually, how to sequence payments to manage intraday liquidity.

Exception Handling

Payment operations generate exceptions: failed transactions, partial matches, formatting errors, returned items. AI can automate much of the investigation and resolution work by learning patterns from historical exceptions. The goal is not full automation but reducing the manual effort required per exception. Results for common exception types are good; unusual cases still benefit from human judgment.

Where Payment Processing Sits on the Maturity Spectrum

Payment optimization is "maturing." The technology works, but implementation complexity varies significantly based on the organization's payment infrastructure. Organizations with standardized, well-integrated payment systems see faster results; those with fragmented legacy infrastructure face longer implementation timelines.

Document Processing and Data Extraction

Treasury operations involve substantial document processing: bank statements, trade confirmations, contracts, invoices, audit requests. AI-powered document processing can reduce the manual effort required to extract and process information from these documents.

While most bank statement data arrives electronically via standardized formats, exceptions persist—PDF statements from smaller banks, confirmation documents, fee schedules that arrive as attachments. And even the so-called "standard" formats such as

SWIFT's or Nacha's (Nacha originally stood for the National Automated Clearinghouse Association), in practice almost always require bank-specific configuration to accommodate each institution's particular conventions. A bank confirmation that previously may have required an analyst to read and key in the relevant details can now be processed automatically, with the confirmation number, trade details, settlement amounts, and dates populated into the TMS for human verification only.

For credit agreements, ISDA masters, and bank fee schedules, AI can identify key terms, flag unusual provisions, and extract structured data—financial covenants, pricing terms, amendment provisions—for tracking and analysis.

Where Data Processing Sits on the Maturity Spectrum

Document processing is "maturing rapidly." For standardized document types in common formats, accuracy is high. For highly variable or proprietary formats, human review remains necessary. The technology has improved significantly in recent years; what required custom development a few years ago is increasingly achievable with available tools.

Natural Language Interfaces

Natural language interfaces—covered in depth in Chapter 2—are finding specific applications in treasury operations. Rather than building a custom report or navigating multiple screens, a treasurer might ask: "Show me all entities where actual cash is more than 20% below forecast this month" or "Which subsidiaries have payments due this week that exceed our single-counterparty limit?"

Beyond answering queries, AI can generate narrative explanations of data. Rather than presenting a forecast chart alone, the system might explain: "Cash is projected to be tight in week three primarily due to the quarterly tax payment coinciding with the semi-annual insurance premium. The forecast assumes normal collections; any delays would increase the shortfall."

Where Natural Language Interfaces Sit on the Maturity Spectrum

Natural language interfaces are "maturing." Straightforward queries work well. Systems still struggle with ambiguity or complex multi-part questions, and users can receive confident-sounding but incorrect answers. The practical approach is to deploy these interfaces for well-defined query types while maintaining traditional reporting for complex analysis.

Foreign Exchange Hedge Management

FX hedge management is one of the most complex and operationally demanding areas of corporate treasury—and one where AI-based approaches show genuine promise, though significant challenges remain. Chapter 7 addresses this in depth. The summary here is intended to give context for where this application sits on the maturity spectrum.

For a global corporation, even identifying and consolidating FX exposures is a substantial challenge. Receivables sit in one module, payables in another, intercompany balances in a third, treasury deals in the TMS. Pulling these together into a coherent, current exposure view requires either manual effort or well-designed integration—and in many organizations, it is primarily the former. By the time exposures are extracted, reviewed, and classified, and a hedge recommendation prepared, the underlying exposures may have already changed.

The intercompany dimension adds further complexity: multilateral netting, entity-specific settlement rules, currency restrictions, and the distinction between cash flow and balance sheet exposures each carry different accounting implications under International Financial Reporting Standards (IFRS) 9 or Accounting Standards Codification (ASC) 815. Managing hedge documentation, effectiveness assessments, and designation memos manually against a reporting deadline is precisely the kind of high-stakes, high-volume administrative work where errors are most likely and consequences most significant.

Where Foreign Exchange Hedge Management Sits on the Maturity Spectrum

FX hedge management is best characterized as maturing across the board. AI can assist meaningfully at multiple points in the workflow—exposure identification, netting calculation, hedge sizing, compliance checking, trade preparation—but the full orchestration of the end-to-end process using a multi-agent architecture is still in early stages. The potential is among the highest of any treasury AI application. The complexity of getting it right is also among the highest. Chapter 7 explores both.

Bank Relationship Management

Treasury maintains relationships with multiple banks, each with its own fee structures, service agreements, and performance characteristics. Bank relationship management is an area where the foundational analytical work is already achievable with current tools—and where AI-based enhancements can add meaningful value on top of that foundation.

Bank Fee Analysis

Bank fee analysis has historically been tedious: fee structures are complex, statement formats vary by bank, and meaningful comparison requires normalizing across different pricing models. The analytical work—parsing fee statements, categorizing charges, identifying anomalies and optimization opportunities—is production-ready in the sense that it can be done reliably with today's tools. An example: analyzing twelve months of fee statements across eight banking relationships to identify that wire transfer fees from one bank have increased 15% without corresponding service improvements, that another bank's lockbox pricing is significantly above alternatives, and that consolidating certain services could reduce overall costs.

Where AI adds value beyond the foundational analysis is in pattern detection at scale and over time. AI can flag fee anomalies as they occur rather than in a quarterly review, identify subtle pricing changes across complex fee structures, and benchmark performance across relationships continuously rather than periodically. These enhancements are becoming increasingly practical.

Regulatory Filings

A related area worth noting concerns the regulatory filings that arise from bank relationships and treasury and finance related operations.

Some examples:

FBAR. The Financial Crimes Enforcement Network (FinCEN) requires US filers to report each foreign account where signature authority or financial interest exists, along with the maximum balance held during the calendar year, using a Foreign Bank Account Report (FBAR), identified as FinCEN Form 114. That information typically lives across bank master data and the bank account management workflow for signatory tracking—in SAP, for example, within the Bank Account Management module—alongside the TMS for account inventory and bank statement archives for balance determination.

Escheatment, the reporting of unclaimed property to state authorities, presents a similar challenge: each US state sets its own dormancy periods, due diligence requirements, filing formats, and deadlines, and the underlying data—uncashed vendor payments, unclaimed payroll, dormant customer balances—sits across accounts payable (AP), payroll, accounts receivable (AR), and banking systems.

AI-assisted aggregation and document processing can help in two ways. The first is pulling the relevant source data from its respective systems into a consolidated working file—signatory records and maximum balances for FBAR, dormancy-aged items by state for escheatment. The second is generating each filing in the format the authority requires, with the analyst's role shifting from compilation to review. The work does not disappear—these filings carry meaningful penalties for error and benefit from human judgment on edge cases—but the time spent on mechanical aggregation can be substantially reduced.

Where Bank Relationship Management Sits on the Maturity Spectrum

Fee analysis is "production-ready" in its foundational form, with AI-based enhancements in the "maturing" category. Service performance monitoring—capturing and analyzing service metrics across banking relationships—is more "experimental," as the data infrastructure to support it often does not yet exist and requires groundwork. Regulatory filing support—FBAR aggregation, escheatment reporting—sits in the maturing category: data extraction and consolidation work well, but the penalty exposure on these filings keeps human review essential.

Reconciliation and Matching

Reconciliation—matching transactions across systems, identifying discrepancies, resolving breaks—consumes significant treasury operations time. AI-based matching can identify likely matches even when exact agreement is absent. The system learns from historical match decisions and applies that learning to new transactions.

A practical example: an incoming payment of $99,847.23 doesn't match any open invoice exactly. Traditional matching fails. The AI system identifies that invoice #12847 for $100,000 is the likely match—the amount difference corresponds to a standard 0.15% early payment discount, and the customer has taken this discount on prior payments.

Where Reconciliation and Matching Sit on the Maturity Spectrum

Intelligent matching is "production-ready" for common reconciliation scenarios. The technology is mature and widely deployed in bank reconciliation and accounts receivable matching. Exception resolution assistance is maturing—helpful for common exception types but still requiring human judgment for unusual cases.

Working Capital Optimization

Working capital management—optimizing the balance of receivables, payables, and inventory—involves complex trade-offs that AI can help evaluate. When to pay suppliers, which overdue receivables to prioritize for collections, how to sequence payments against the cash forecast—AI can analyze these trade-offs across thousands of relationships simultaneously, considering discount terms, cash position, relationship importance, and historical patterns.

Where Capital Optimization Sits on the Maturity Spectrum

Working capital optimization is "maturing." The analytical capabilities exist, but implementation requires integration across AP, AR, and treasury systems that many organizations are still building. Where the data infrastructure exists, results can be significant; building that infrastructure is typically the primary challenge.

What's Still Experimental

Several applications generate significant discussion but are not yet ready for unsupervised production deployment. These were covered in Chapter 2, but here, I am using treasury examples.

Autonomous decision-making. AI systems that make and execute decisions without human approval—automatically hedging exposures, repositioning cash, adjusting investment allocations—remain largely experimental. The technology is capable in narrow domains, but organizational readiness, control frameworks, and regulatory clarity are still developing alongside it.

Strategic recommendations. AI can support analysis effectively, but relying on it for strategic decisions—capital structure, financing structure, long-term planning—without significant human oversight is premature. These decisions involve judgment, context, and stakeholder considerations that current AI handles poorly.

Complex negotiation support. Using AI to support negotiations around credit agreements, banking arrangements, or counterparty contracts is emerging but not mature. The nuance required in these interactions exceeds current AI capabilities.

Patterns Across the Landscape

Looking across these applications, several patterns are worth noting.

Data quality is the binding constraint. In nearly every domain, the readiness of data is the primary limiter, rather than the capability of the AI. Organizations that have invested in clean, consistent, accessible data are meaningfully better positioned than those that have not.

Human oversight remains essential. Even in production-ready applications, human review is part of the process. AI augments human judgment rather than replacing it. The question is where in the workflow human involvement is most valuable—not whether it is needed at all.

Integration complexity varies. Some applications can be implemented relatively independently. Others require integration across multiple systems. Implementation timelines and costs vary accordingly, and this is worth understanding before committing to a particular initiative.

The pace of development is rapid. Capabilities that were experimental two years ago are maturing today. The landscape is evolving quickly, making it worth reassessing periodically rather than forming fixed views.

What's Next: From Analysis to Action

The applications surveyed in this chapter share a common characteristic: they analyze and recommend. A human reviews the output and decides what to do. This is valuable and is where most organizations should start. But it is not where the technology stops.

The next frontier is AI that doesn't just analyze—it acts. AI systems that observe a cash position and execute the sweep. That detect fraud patterns and hold the payment. That calculate exposures and initiate the hedge. These are AI agents: systems that can take action within boundaries you define, not just present recommendations for human approval.

The shift from analysis to action raises the stakes. When AI only recommends, a wrong answer can be caught by human review. When AI acts, the consequences are immediate. This demands a different architecture—one designed for safety, auditability, and controlled autonomy. That architecture is what we explore in Chapter 4.

PUT IT INTO PRACTICE

Assess your data readiness for the treasury AI applications most relevant to your organization.

For each major treasury process, ask the following questions:

- Is historical data available in a structured, accessible format?
- How far back does reliable, consistent data extend?
- Have definitions, categorizations, or systems changed in ways that create inconsistencies in that history?
- Could you extract this data for analysis if needed, and who would need to be involved?

This is not a comprehensive data audit—just an honest initial assessment. The answers will help you identify where an AI initiative might be straightforward versus where data groundwork needs to come first. Most organizations find that data readiness varies significantly across processes, and that variation is itself useful information for prioritization.

CHAPTER SUMMARY

- Cash forecasting is maturing rapidly—the potential is genuine and the proof-of-concept work is encouraging, but many organizations still rely on spreadsheets and manual processes; Chapters 5 and 6 explore this in depth.
- Fraud detection is production-ready and widely deployed, using pattern recognition to identify anomalous transactions; it complements rather than replaces fundamental controls.
- Payment processing optimization is maturing, with good results for routing, timing, and exception handling in well-integrated environments.
- Document processing is maturing rapidly, with high accuracy for standardized formats and ongoing human review for unusual ones.

- Natural language interfaces are maturing, practical for well-defined queries, still developing for complex or ambiguous ones.
- FX hedge management is maturing—this area is complex, with high stakes and a lot of potential. Chapter 7 addresses the full process and the multi-agent architecture that applies to it.
- Bank relationship management is production-ready in its foundational form, with AI enhancements maturing; service performance monitoring is more experimental.
- Intelligent matching is production-ready for common reconciliation scenarios.
- Working capital optimization is maturing but may be constrained by data integration requirements.
- Autonomous decision-making remains experimental across all treasury domains.
- Data quality is consistently the binding constraint—AI capability typically exceeds data readiness in many organizations.

In the next chapter, we'll examine the agent architecture—what AI agents are, how they work together, and why this approach is essential for safe, auditable automation in treasury.

CHAPTER 4

The Agent Architecture

How AI Systems Can Work Together Safely for Treasury

A single AI system that tries to do everything is difficult to trust, audit, and maintain. This chapter introduces the multi-agent architecture—an orchestrator coordinating specialized sub-agents, each responsible for one thing—and explains why this design is the right foundation for AI in a treasury environment, where accountability and auditability are non-negotiable.

So far, we've discussed AI in terms of what it can do—recognize patterns, make predictions, understand language, generate recommendations. But we haven't addressed a fundamental question: how do you structure an AI system so that it can actually take action in an enterprise environment, where the stakes are real, the data is regulated, and a wrong decision could result in a mispriced hedge or a failed audit?

This chapter attempts to answer that question. We'll start with the concept of AI agents—what they are and why they matter. Then we'll explore why the most effective enterprise implementations use multiple specialized agents working together, rather than a single all-purpose system. Along the way, we'll walk through concrete treasury examples to make these concepts tangible.

What Is an AI Agent?

An AI agent is an AI system that can take action, not just provide analysis.

This distinction is fundamental. Most ML applications I have worked with are passive—they analyze data and present results for humans to act upon. A dashboard shows your cash position. A forecast predicts next week's flows. A fraud detection system flags suspicious transactions for review.

An AI agent goes further. It observes a situation, decides what to do, and executes that decision—all within boundaries you define.

Consider the difference in a treasury context:

Passive AI: "Based on current positions and forecast, I recommend drawing $5M from the revolver to cover the projected shortfall on Thursday."

AI agent: Observes the projected shortfall, evaluates it against your liquidity policy, determines that a revolver draw is appropriate, initiates the draw request, and notifies you that it's been done.

The passive AI informs. The agent acts.

This shift from informing to acting is what makes agents both powerful and—if poorly designed—potentially risky. When AI only recommends, a human reviews every decision. When AI acts, you need different safeguards.

Why Agents Matter for Treasury

Treasury operations involve countless decisions that follow well-defined logic. When the balance in a concentration account exceeds a threshold, sweep the excess. When an FX exposure reaches a policy limit, initiate a hedge. When a payment matches fraud patterns, hold it for review.

Today, humans make most of these decisions—often reviewing system recommendations and clicking "approve." This works, but it creates bottlenecks. As AI systems get smarter and generate more recommendations, the human becomes the constraint. You can't scale human attention the way you can scale computation.

Agents offer a path forward: delegate routine decisions to AI systems that can act within defined boundaries, while humans focus on exceptions, strategy, and judgment calls that genuinely require human insight. But this only works if the agents are designed correctly—which brings us to the architecture question.

Why Single-System Architecture Falls Short

When organizations first consider AI agents, the instinct is often to build one comprehensive system. "Let's create an AI that handles our entire FX exposure process—from extracting positions to calculating exposures to recommending hedges to executing trades."

This seems logical. Why fragment the work when AI is so capable? The answer lies in understanding how complex systems fail—and what happens when they do.

When Monolithic Systems Go Wrong

Imagine a single AI system that handles your entire FX workflow. One day, it produces a hedge recommendation that looks wrong. Where do you start investigating?

Was the underlying data extracted incorrectly? Was there an error in the netting calculation? Did the hedge sizing logic malfunction? Was the counterparty selection flawed? When everything is intertwined in one system, diagnosing problems requires examining the entire process.

Now imagine the auditor's question: "Why did the system recommend this specific trade?" In a monolithic system, the answer is essentially "the AI decided"—not a satisfying response for regulatory purposes.

And consider what happens when policy changes. Your hedge ratio moves from 75% to 80%. In a monolithic system, that logic might be embedded in multiple places, requiring careful surgery to update without breaking something else.

It is worth noting that monolithic architecture is common among commercial AI platforms and off-the-shelf AI tools, precisely because it is simpler to deploy as a single system. That simplicity is a reasonable trade-off in many contexts. In treasury, where regulatory explainability, Sarbanes–Oxley Act (SOX) audit trails, and the ability to change policy parameters without touching the underlying logic are not optional, it is a trade-off worth understanding clearly before choosing a platform or commissioning a build.

The Alternative: Sub-Agents and the Orchestrator

The more robust approach is to break the work into specialized sub-agents, each handling one well-defined piece of the process, coordinated by an overarching controlling intelligence called the orchestrator.

The *orchestrator*—think of it as the master controller—receives the overall objective, breaks it into discrete tasks, assigns each task to the appropriate sub-agent, validates the output before passing it forward, and enforces the policy rules that govern the entire pipeline. The orchestrator does not do the work itself. It directs, sequences, validates, and controls.

The sub-agents are the specialists. Each one receives a clearly defined input, performs exactly one function, and returns a structured output. A sub-agent that extracts intercompany transaction data from the enterprise resource planning (ERP) system knows nothing about netting logic, hedge sizing, or compliance rules. It extracts data and passes it on. That's all.

The relationship between the orchestrator and the sub-agents reflects something treasury professionals will recognize immediately: it is the same principle that underpins the segregation of duties in manual treasury operations. Just as no single individual should have authority over every stage of a transaction—from initiation to approval to execution—no single agent should control the full pipeline. Authority is distributed, each party is accountable for their specific function, and the controlling intelligence—the orchestrator, in the role of the treasury manager—oversees the whole.

In practice, for an FX hedge management use case, this architecture looks like five sub-agents working in sequence:

- One sub-agent extracts intercompany transaction data from the ERP system.
- A second calculates net positions and identifies the FX exposures created at the in-house bank level.
- A third generates specific hedge recommendations for each material exposure.
- A fourth checks each recommendation against policy limits, approved counterparties, and compliance rules.
- A fifth handles execution—posting approved trades to the TMS and generating confirmations.

Each sub-agent receives clean inputs and produces structured outputs. The orchestrator sequences the work and validates results at each handoff. Nothing passes to the next stage without the orchestrator's validation check clearing.

How Multi-Agent Systems Work

Think of a well-run SAP treasury implementation (easier said than done!). The solution architect or project lead receives the project scope, defines the workstreams, and assigns each to specialists. One team handles functional design—mapping business requirements to system configuration. Another manages the technical build and integration, connecting SAP to the bank connectivity layer, the ERP system, and any surrounding systems. A third focuses on controls, governance, and compliance—ensuring the design meets audit and regulatory requirements. A fourth runs testing, and a fifth handles documentation and training. No single team does everything. Each works within a defined scope and hands off its outputs to the next, and the solution architect ensures the sequencing holds together

The solution architect doesn't try to do everything personally—but they control the sequence, review each deliverable before it passes to the next team, and ensure the final output meets standards.

A multi-agent AI system works exactly this way. The orchestrator is the solution architect. The sub-agents are the expert teams—each focused on one well-defined task, receiving clear inputs and producing structured outputs.

The Single Most Important Principle

Each sub-agent does exactly one thing and passes a clean, structured output to the next sub-agent.

This isn't a stylistic preference. It is the architectural expression of a control principle that treasury professionals already apply in their daily operations: segregation of duties.

In manual treasury processes, segregation of duties means that the person who initiates a payment is not the person who approves it. The person who enters a deal is not the person who confirms it with the counterparty. The person who reconciles the position is not the person who executed the trade. These separations exist because no single individual should have unchecked authority over a complete transaction cycle—and because accountability is clearest when each person's responsibility is clearly defined.

Applied to sub-agents, the same principle means that the sub-agent which generates a hedge recommendation should not be the sub-agent that checks compliance. The sub-agent that calculates net positions should not be the sub-agent that executes trades. Each sub-agent is authorized to perform only the specific function it was designed for—nothing more. The orchestrator enforces this boundary, just as a treasury manager enforces role boundaries in a manual process.

When something goes wrong—and in complex systems, things occasionally will—each handoff between sub-agents is a checkpoint. If the netting calculation is wrong, you examine the netting sub-agent. If a compliance check fails, you know exactly which check it was and what triggered it. This diagnostic clarity is essential for audit trails. When regulators ask, "Why did the system make this decision?", a multi-agent architecture provides a clear answer: "Sub-agent 2 produced these net positions. Sub-agent 3 applied these rules to generate this recommendation. Sub-agent 4 checked these compliance criteria. Here are the inputs and outputs at each stage."

A Concrete Example: Foreign Exchange Exposure Consolidation

To make these concepts tangible, let's walk through how a multi-agent system handles FX exposure consolidation through an in-house bank (IHB) netting structure. (An IHB is a central treasury entity that acts as an internal bank for subsidiaries.)

This use case is deliberately challenging. FX intercompany netting involves multiple data sources, cross-currency mechanics, regulatory constraints (IFRS 9 hedge accounting), audit requirements (SOX controls), and a clear business outcome (hedge recommendations that can be executed). If an architecture can handle this correctly, it can handle most treasury workflows.

The Business Context

Many multinational organizations use in-house banks to centralize intercompany settlements. Instead of each subsidiary settling individually with each counterpart generating dozens of cross-border payments and currency conversions—all intercompany positions flow through a central IHB. The IHB acts as the counterparty to all subsidiaries. Each subsidiary has a single current account at the IHB denominated in its functional currency. On netting date, each subsidiary receives a single debit or credit representing the net of all its intercompany positions. This shift from bilateral to multilateral netting is arguably one of the most significant structural efficiency gains available to a multinational treasury.

This structure dramatically reduces settlement costs and concentrates FX risk in one place—the IHB—where it can be hedged efficiently. The complexity arises when subsidiaries have obligations in currencies different from their functional currency. A UK subsidiary (GBP functional) that owes EUR to the IHB creates an FX exposure: the IHB receives GBP but needs EUR to balance its clearing accounts. This exposure must be identified, quantified, and hedged.

The Five Sub-Agent Pipeline

Here is how the architecture could work with five sub-agents:

Sub-agent 1: data extraction. The first sub-agent retrieves intercompany transaction data from the ERP system. It receives parameters (which entities, what date range) and returns a structured list of intercompany line items—amounts, currencies, counterparties, due dates. This sub-agent knows nothing about netting logic, hedge sizing, or compliance rules. It extracts data and passes it on.

Sub-agent 2: netting engine. The second sub-agent applies multilateral netting logic. It receives the raw intercompany transactions and calculates net positions—what each entity owes or is owed, by currency, after all intercompany positions cancel against each other. Crucially, it also identifies the FX exposures created at the IHB level. The netting engine is where functional treasury expertise is most directly encoded: the logic for multilateral netting, functional currency handling, and IHB exposure calculation comes from treasury knowledge—the sub-agent executes that logic consistently at scale.

Sub-agent 3: trade processing. The third sub-agent receives the IHB's net FX exposures and generates specific hedge recommendations. For each material exposure, it produces a complete trade specification: instrument type (FX forward, FX option), notional amount, currency pair and direction, tenor, counterparty selection. The recommendations include written rationales—why this instrument, why this amount, why this counterparty. This is the audit trail that supports hedge accounting documentation and satisfies control requirements.

Sub-agent 4: approval and compliance. The fourth sub-agent checks each recommendation against policy limits and compliance rules: trade size limits, approved counterparty lists, instrument restrictions, credit line availability—all the constraints that a treasury policy manual would specify. Recommendations that pass all checks move forward. Recommendations that exceed thresholds are escalated to appropriate approvers. Recommendations that violate hard constraints are blocked.

Sub-agent 5: execution. The fifth sub-agent handles approved trades—posting them to the treasury management system, generating confirmations, updating position records. In a fully autonomous implementation, this sub-agent executes without human intervention for trades within pre-approved parameters. In a more conservative implementation, it prepares everything for one-click human execution.

This five-agent example illustrates the architecture principle. In practice, a production treasury AI system will involve significantly more agents—and more complex logic within each. The principle of single responsibility holds regardless of scale.

The Orchestrator: Where Policy Lives

Above all five sub-agents sits the orchestrator—the controlling intelligence that sequences work, validates outputs, and enforces policy rules.

The orchestrator is not the most technically complex component. It is the most strategically important—because it is where treasury policy becomes executable logic. Think of the orchestrator as your standing instructions to the entire system. Every sub-agent operates under these instructions. Every check runs against these rules. When policy changes, updates propagate from one place rather than requiring modifications throughout the system.

The orchestrator also enforces the segregation of duties between sub-agents at a structural level. It ensures that the sub-agent responsible for generating a recommendation is never the same sub-agent that validates it—the architectural equivalent of the four-eyes principle that governs manual treasury authorization. This is not redundancy; it is control design.

Guardrails Between Sub-Agents

The orchestrator's most critical operational function is running validation checks at each handoff between sub-agents. These checks act as gates—nothing passes to the next stage without validation.

For the FX example, checks between the netting engine and trade processing might include the following: data quality verification (did the netting calculation complete successfully and are all expected entities represented?), rate freshness (are the spot rates current?), variance checks (are the calculated exposures within reasonable range of prior periods?), and materiality screening (which exposures are large enough to warrant hedging?).

Checks before execution might include the following: policy limits (does any single trade exceed size thresholds requiring escalation?), counterparty constraints (are recommendations limited to approved counterparties?), instrument restrictions (are only permitted hedge instruments being recommended?), and credit availability (is there sufficient credit line with the selected counterparty?).

Some checks produce warnings—flags for human review before proceeding. Some produce escalations—routing to appropriate approvers. Some produce hard stops—the pipeline halts entirely until the issue is resolved.

Why Guardrails Sit Between Sub-Agents, Not Inside Them

This architecture point deserves emphasis. Guardrails sit between sub-agents, in the orchestrator's validation layer—not inside the sub-agents themselves. This is what makes the system maintainable. If your hedge policy changes from 75% coverage to 80%, you update one parameter in the orchestrator. If you add a new approved counterparty, you update one list. Each sub-agent remains unchanged. The single-responsibility principle applied to policy management means that policy changes are surgical rather than scattered.

The Business Value of This Architecture

The business value of the orchestrator and sub-agents architecture lies in its auditability, transparency, maintainability, and (controlled) autonomy.

Auditability. Every handoff between sub-agents is documented. Inputs, outputs, timestamps, validation results—all captured. When audit asks, "Why did the system recommend this trade?", the answer is traceable through each stage.

Transparency. Because each sub-agent has a single responsibility, its behavior is understandable. The netting engine calculates nets. The trade processor sizes hedges. The compliance checker validates against policy. No mysterious black-box decision-making.

Maintainability. Policy changes, new requirements, regulatory updates—all can be implemented by modifying specific components rather than rebuilding the entire system.

Controlled autonomy. The architecture supports graduated autonomy. Initially, the system might stop at recommendations, requiring human approval for each trade. As confidence builds, autonomy can expand—auto-executing trades below certain thresholds, with human approval only for exceptions. The guardrail system makes this graduation safe. At each stage, controls can be adjusted to match organizational comfort level.

A Practical Observation

The FX exposure system described in this chapter represents a pattern that applies broadly across treasury—and across finance more generally.

What makes it work isn't the sophistication of any individual sub-agent. It's the architecture: clear separation of responsibilities, structured handoffs, validation at each stage, centralized policy management. These are the same principles that make manual treasury operations trustworthy—segregation of duties, authorization hierarchies, audit trails—expressed in the design of an AI system.

This is good news for treasury professionals. You don't need to understand deep learning or neural network architectures to design effective AI systems. You need to understand your domain: which data matters, which rules apply, which exceptions require escalation, what audit trail regulators require. That domain expertise, translated into clear sub-agent specifications and orchestrator policy rules, is what makes an AI treasury system production-grade rather than a prototype. The technology is a tool. The expertise is the product.

PUT IT INTO PRACTICE

Map one of your complex treasury workflows as a sub-agent pipeline.

Choose a workflow you know well—cash positioning, FX exposure management, payment processing, or intercompany settlement.

Then sketch out:

- What would each sub-agent do? (One responsibility per sub-agent, strictly enforced.)
- What inputs would each sub-agent receive?
- What outputs would each sub-agent produce?
- What checks would the orchestrator run between stages?
- What policy rules need to be enforced—and where would segregation of duties require that the sub-agent generating a recommendation be different from the one validating it?

Understanding your workflow this way is the first step toward designing an AI system that could support it—and toward recognizing that the control principles you already apply in manual treasury operations are the same principles that make AI systems trustworthy.

CHAPTER SUMMARY

- An AI agent is an AI system that can take action, not just provide analysis—it observes, decides, and executes within defined boundaries.
- Monolithic AI systems are difficult to audit, diagnose, and maintain; when they fail, the failure is hard to trace, and policy changes require surgery throughout the system
- The multi-agent architecture uses an orchestrator and specialized sub-agents: the orchestrator controls, sequences, and validates; each sub-agent performs exactly one main function.
- Segregation of duties—a fundamental principle of manual treasury control—applies directly to sub-agents: no single sub-agent should have authority over a complete transaction cycle, and the sub-agent that generates a recommendation should never be the one that validates it.
- Guardrails sit between sub-agents in the orchestrator's validation layer, not inside the sub-agents themselves—making policy changes surgical and creating clear audit checkpoints.
- The FX netting pipeline example demonstrates the following architecture: data extraction → netting engine → trade processing → compliance check → execution. Each part is a separate sub-agent with defined inputs, outputs, and authority.
- Domain expertise is the key asset: clear sub-agent specifications and orchestrator policy rules, built on treasury knowledge, are what make the difference between a prototype and a production system
- The architecture supports graduated autonomy—starting with human approval for all decisions, then selectively expanding as confidence in the system is established.

In the next chapter, we'll start PART II by examining intelligent cash forecasting in detail—starting with the problem itself, and then exploring what AI-based approaches might do about it.

PART II

The Architecture

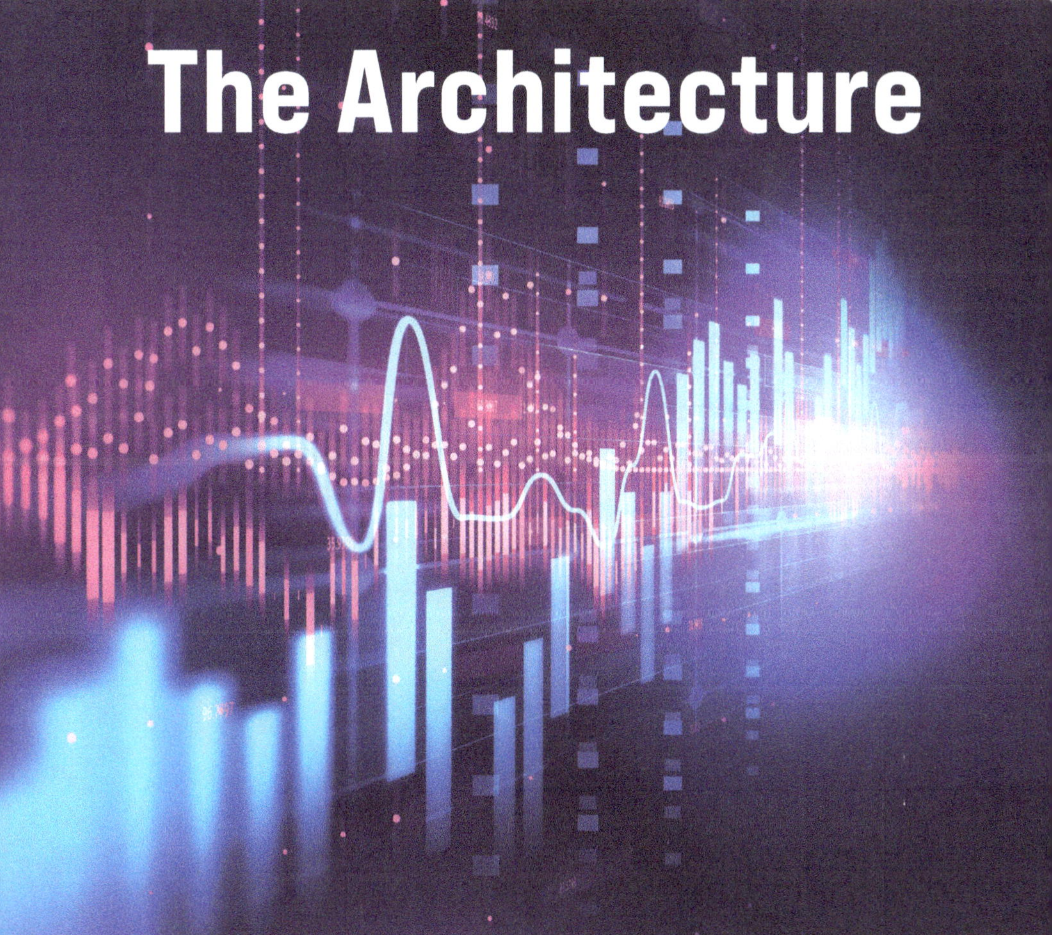

> **"It is far better to foresee even without certainty than not to foresee at all."**
>
> **Henri Poincaré**
> The Foundations of Science (1913)

CHAPTER 5

The Cash Forecasting Engine

The Cash Forecasting Challenge—and What AI Might Do About It

Cash forecasting is one of treasury's most persistent and underappreciated challenges—and the accountability structure around it matters as much as the technology. This chapter explores why the problem is genuinely hard, what good forecasting actually means in practice, and how an ensemble AI architecture can be designed to address it. It also addresses the accountability dimension directly: AI can identify variance causes faster, and shared organizational ownership of forecast accuracy is what makes the improvement durable.

Of all the routine challenges in corporate treasury, cash forecasting is the one I have seen cause the most consistent frustration across the most organizations. Not because treasury professionals are not good at it—they are, and they work hard at it—but because the problem itself is genuinely difficult, the tools available have historically been inadequate for the task, and the gap between a forecast that "kind of works" and one that actually improves decisions is wider than most people outside treasury appreciate.

I want to start this chapter with the problem, not the solution. There is a tendency in any discussion of AI to be drawn toward the technology—the models, the algorithms, the architecture. But the technology is only worth exploring once the problem it is meant to address is clearly understood. Also, I recommend beginning AI implementation with a clearly defined area of focus (more on this in Chapter 11). So before we get to AI, let us spend some time on why it is as hard as it is, and what "good" could actually look like.

The Universal Pain

Ask any corporate treasurer about their cash forecasting process, and you will get one of two responses. The first is a slightly rueful acknowledgment that the process works—just not as well as it should, for reasons that everyone understands but nobody has quite managed to fix. The second is a more direct expression of frustration: the numbers are consistently wrong, the process consumes more time than it should, and the downstream decisions made on the basis of those numbers are consequently less reliable than they need to be.

In my experience, both responses are telling the same story. The first reflects an organization that has learned to live with imperfect forecasting by building buffers, maintaining larger-than-necessary cash reserves, and making conservative decisions to compensate for uncertainty. The second reflects an organization where the dysfunction is more visible, but the underlying causes are the same.

Walk into almost any corporate treasury center, and you will find a version of the same scene. A spreadsheet—sometimes several spreadsheets, sometimes a labyrinthine network of spreadsheets linked by formulas that only one person fully understands—sits at the center of the daily cash positioning process. Subsidiaries have been asked to submit their forecasts by a certain time. Some have. Others have not. The ones who have submitted have used their own formats, their own categorizations, their own interpretations of what "forecast" means.

By the time the daily position is assembled, distributed, and reviewed, some of what it contains is already out of date. Decisions that needed to be made at 9 am are being made at 11 am, on information that reflects yesterday's reality more than today's. And when tomorrow's forecast varies from what actually occurred—as periodically it will—nobody is quite sure whether the gap came from the AR estimate, the intercompany netting, the subsidiary that submitted late, or the payment that cleared two days earlier than expected.

This is not a description of poorly run treasury operations. It is a description of how treasury forecasting works in many sophisticated, well-resourced multinational organizations. The problem is not the people. It is the process and the tools.

Why It Is Harder than It Looks

Cash forecasting sounds deceptively simple. You are trying to predict how much money will be in your accounts on future dates. The data is, in principle, available: transaction history, open invoices, payment schedules, bank balances. How hard can it be?

In practice, several things make it genuinely difficult—and understanding these is the starting point for understanding where AI-based approaches might help.

The Interaction Challenge

Cash flows are not driven by a single variable. They are driven by dozens of variables that interact with each other in ways that are often non-obvious. Seasonal revenue patterns interact with customer payment terms. Billing cycle timing interacts with month-end close processes. Payroll schedules interact with tax payment obligations. Capital expenditure programs interact with debt service schedules.

The challenge is not that any one of these is hard to model. It is that the combinations multiply rapidly, and the interactions produce outcomes that no individual analyst can reliably anticipate. A seasonal revenue peak that coincides with a quarterly tax payment and a large supplier invoice produces a very different cash outcome than any of those three factors in isolation.

The Horizon Dimension

Different forecast horizons are fundamentally different problems, not just longer or shorter versions of the same thing. A T+1 forecast is essentially an extrapolation of very recent activity, heavily dependent on what cleared today and what is confirmed to clear tomorrow.

A T+30 forecast is a structural pattern recognition problem: what does this month historically look like, what scheduled events are approaching, what seasonal factors apply?

The forecasting approach that works well at one horizon frequently fails at another. The analyst who is excellent at calling the daily position—because they know the business intimately and can track individual large transactions—is often less reliable at the monthly view, where the sheer number of variables overwhelms individual judgment.

The Data Dimension

Before any forecast can be built, someone has to reckon with the data. And in most organizations, the data available for cash forecasting is messier than it appears. Categorizations change over time as the chart of accounts structures evolves. Subsidiaries interpret the same categories differently. Large one-off items—acquisitions, settlements, restructuring charges—distort historical patterns if not handled carefully. A forecast built on inconsistent historical data will inherit those inconsistencies, producing errors that are difficult to diagnose because they are baked into the foundation rather than visible in the model.

The Accountability Problem

Perhaps the most underappreciated dimension of the cash forecasting challenge is the accountability structure around it. The people who provide inputs to the forecast—the AR team estimating collections, the AP team projecting disbursements, the business units submitting local forecasts—are not the people who suffer the consequences when the forecast is wrong. The treasury team bears the operational risk of forecast error without controlling the inputs that drive it.

No AI system can fix a broken accountability structure. But understanding that this structure exists—and that improving it is a prerequisite for any forecasting improvement, technical or otherwise—is important context for what follows.

What changes this dynamic, in practice, is senior leadership engagement. In organizations where the CFO has made cash forecast accuracy a shared organizational responsibility—not just a treasury metric—the quality of forecast inputs improves measurably. When the people providing forecast data understand that variances will be traced back to their inputs, and that those variances will be discussed at a senior level, the discipline around submitting accurate, timely numbers changes. The accountability structure shifts from one-sided to genuinely shared.

The mechanism is variance analysis. In practice, this is where the accountability lever actually sits. When a forecast variance occurs, the question is not only "how large was it" but "what caused it." A simple example illustrates the point: in one organization, the CFO could not understand why accounts payable was difficult to forecast with precision when standard payment terms meant the liability was known thirty days in advance. The answer, surfaced through variance analysis, was that forty percent of invoices were being entered into the AP system the day before payment was due—making it impossible for treasury to incorporate them into any meaningful forecast. That finding prompted a direct operational change in how the AP team processed invoices. Treasury had not just improved its forecast; it had helped the organization operate better. There is a broader benefit worth naming too: when treasury works collaboratively with the departments that provide forecast inputs—helping them understand how their numbers flow into decisions and where variance originates—those departments often find it improves their own planning as well. The relationship that starts as an accountability conversation tends to become a more productive working partnership over time.

This is the practical case for AI-assisted forecasting that goes beyond accuracy metrics. The faster treasury can identify and explain forecast variances—by category, by entity, by the specific input that drove the deviation—the faster that information can reach the people who can act on it. In most treasury teams today, assembling a variance explanation takes the better part of a day. A team that can produce a clear, category-level variance analysis by mid-morning is not just better at forecasting; it is better positioned to hold the rest of the organization accountable for the inputs that make forecasting possible in the first place. AI does not solve the accountability problem. But it removes the excuse that the problem cannot be diagnosed quickly enough to act on.

What "Good" Would Actually Look Like

A treasurer who manages cash well does not need a perfect forecast—and no forecast ever will be. A forecast is by its nature forward-looking: an informed estimate of what has not yet happened, not a guarantee of what will. The goal is not perfection but a forecast that is accurate enough, current enough, and transparent enough to support confident decision-making.

Accurate enough means that the uncertainty in the forecast is smaller than the decisions it is informing. If you are deciding whether to draw on a revolving credit facility, you need to know whether your projected shortfall is real or an artifact of a conservative subsidiary estimate.

Current enough means that the forecast reflects today's reality, not yesterday's. A daily cash position assembled at 11 am from data that stopped updating at 6 pm is not a live position—it is a snapshot from the past.

Transparent enough means that when the forecast is beyond acceptable accuracy percentages, and the forecast diverges from actual outcomes—as periodically it will—the reasoning behind it is visible. A forecast that says "your closing balance will be $4.2 million" is less useful than one that also explains what is driving that number at the category level—and that can be challenged, refined, and supplemented when the treasurer has information the model does not. Transparency is also what enables the continuous improvement that turns a reasonable first forecast into a consistently reliable one.

The Morning Briefing Problem

There is a specific manifestation of all these challenges that captures the daily reality of treasury forecasting better than any abstract description. Every treasury professional who manages a daily cash position knows the experience. Before you can make the morning's funding and investment decisions, you need to know where you stand. In most organizations, assembling this information takes the better part of the first hour of the day—pulling bank statements, checking the TMS, chasing subsidiaries that have not yet submitted, reconciling numbers that do not quite agree across systems.

By the time the position is assembled, some of the decisions it was meant to inform have already been made on partial information, or have been delayed waiting for the position to be ready. The morning briefing that should take ten minutes has taken sixty. And the position that was assembled—at some cost in time and effort—will be out of date again before the afternoon's decisions need to be made.

This is the problem that a well-designed AI-assisted forecasting system should address first. Not the exotic challenge of predicting cash flows three months out with perfect accuracy. The immediate, daily, operationally consequential problem of having a reliable, current, explainable view of today's position available at the start of the working day—and updated continuously as the day progresses.

Introducing AI-Based Approaches—What They Might Offer

With the problem clearly in view, it becomes easier to assess what AI-based approaches might realistically contribute. The honest answer is this: potentially quite a lot, in some areas—with significant caveats about what remains unproven.

Machine learning excels at exactly the kind of multi-variable pattern recognition that makes cash forecasting hard for humans. An ML model can identify patterns across dozens of variables simultaneously—seasonal effects, payment term interactions, billing cycle timing, proximity to key dates—including interactions that no analyst explicitly programmed and that no rule-based system could anticipate.

AI can also address the horizon problem directly, by using different models for different horizons and combining them. This is the concept of an ensemble—a term worth understanding because it is central to how the forecasting approach described in Chapter 6 is designed.

An ensemble, in the context of AI forecasting, simply means combining the outputs of multiple models rather than relying on any single one. Think of it like asking several experienced forecasters for their views and then weighting their inputs based on where each has historically been most reliable. In treasury forecasting, different models excel at different time horizons. An ensemble exploits that by letting each model contribute where it is strongest.

Three model families I use are particularly relevant for treasury time-series forecasting:

ARIMA (auto-regressive integrated moving average) identifies how much of today's cash flow can be explained by yesterday's, and the day before, and the trend between them. It appears best suited to short-horizon forecasting, especially daily cash positioning—one to three days—where recent history is genuinely informative.

XGBoost (extreme gradient boosting) can handle a rich feature set—day of week, proximity to payroll dates, customer payment patterns, invoice aging—and identify non-linear interactions that simpler models miss. It appears best suited to medium-horizon forecasting, roughly one to three weeks

Prophet handles trend, seasonal cycles, and the effects of known events—payroll dates, tax deadlines, debt service. It appears best suited to longer-horizon forecasting, four weeks and beyond, where structural patterns dominate over recent noise.

The critical observation—and one that the proof-of-concept work in Chapter 6 was specifically designed to test—is that no single model performs best across all horizons. Figure 5.1 illustrates this directly, showing how accuracy changes across the four forecast horizons in the POC exploration.

FIGURE 5.1 illustrates the core argument for ensemble design: different models are needed at different horizons.

Where Human Judgment Remains Essential

Before moving on, it is worth being direct about what AI-based forecasting approaches cannot address, regardless of how sophisticated the models become.

AI cannot fix broken data. A model trained on inconsistent historical data will learn the inconsistencies. Data quality is a prerequisite, not a problem that AI solves.

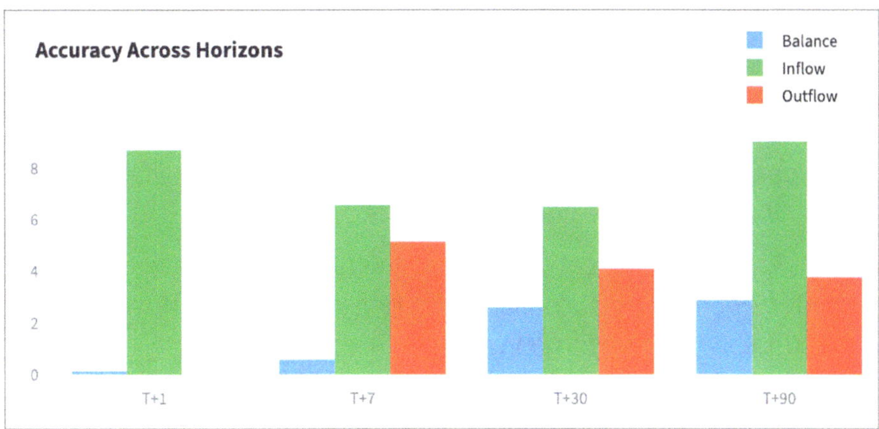

FIGURE 5.1: **Accuracy comparison across T+1, T+7, T+30, and T+90 horizons from the proof-of-concept, showing how balance error (blue), inflow error (green), and outflow error (red) increase with horizon length. This illustrates the core argument for ensemble design: different models are needed at different horizons.**

Sample data only.

AI cannot replace the treasury professional's judgment. The model does not know that a major customer called yesterday to say their payment will be delayed. It does not know that the board approved an unplanned capital expenditure last week. The treasury professional knows these things, and the ability to incorporate that knowledge through overrides is not a weakness in the design. It is the design.

AI cannot make the accountability structure work. If subsidiaries submit late, if business units forecast conservatively, if data governance is inconsistent—these are organizational problems that require organizational solutions. Part of that design involves identifying which decisions require meaningful human review, and building the review points into the system architecture—covered in Chapters 9 and 10. With those caveats clearly stated, the potential that remains is real and worth exploring carefully. That exploration is what Chapter 6 describes.

Chapter 6 describes what happened when this architecture was actually built and tested on real treasury data patterns—including what the results looked like at each forecast horizon, and what the experience revealed about the gap between architectural soundness and production readiness.

PUT IT INTO PRACTICE

Before investing in any forecasting system or technology, understand what you are actually improving on. For your primary operating account, answer these questions honestly:

- What is your current forecasting method? Manual spreadsheet, TMS-generated, subsidiary submissions, some combination? Who owns each component?
- How long does it take to assemble the daily cash position each morning? What are the three steps that consume the most time?
- Do you currently measure forecast accuracy? If not, that is your first deliverable. You cannot know whether any change is an improvement without a baseline.
- What is the most common reason your forecasts miss? Subsidiary non-submission? Unexpected timing of large transactions? Categorization inconsistencies?
- How far back does your clean, consistent cash flow history go? Two years? Five? Less than one?
- Map your forecast accountability structure. For each major category of cash flow—AR collections, AP disbursements, payroll, intercompany—identify who provides the forecast input, how reliably they submit it, and what your variance history shows about whose numbers tend to be most and least accurate. This map is the diagnostic that makes a productive conversation with a CFO possible. You do not need AI to have that conversation—but you do need the data.

CHAPTER SUMMARY

- Cash forecasting is one of treasury's most consistent pain points—not because of poor execution, but because the problem is genuinely hard: too many interacting variables, multiple horizon types requiring different approaches, inconsistent source data, and an accountability structure that works against accuracy.
- The morning briefing problem—the daily effort of assembling a reliable, current cash position before making funding and investment decisions—captures the practical cost of inadequate forecasting tools better than any abstract description.
- Good forecasting means accurate enough, current enough, and transparent enough to support confident decisions—with explainability that allows the treasurer to challenge, verify, and override the model.
- An ensemble—combining multiple models weighted by horizon—is the architectural approach the proof of concept explored; ARIMA for short-term daily cash positioning, XGBoost for the medium term (one to three weeks), and Prophet for the medium to long term (four weeks and more).
- No single model performs best across all horizons—the accuracy chart shows this directly; the ensemble design exploits each model's strength at the horizon where it is most reliable.
- AI does not solve the accountability problem in cash forecasting—but it removes the most common excuse for not acting on it. When variance causes can be identified and explained at a category level by mid-morning rather than end of day, treasury has the diagnostic speed needed to hold forecast contributors accountable and to make the case to senior leadership for shared organizational ownership of forecast accuracy.
- AI cannot fix broken data, replace treasury judgment, or repair a broken accountability structure—these are prerequisites, not problems that technology solves.

In the next chapter, we turn to what an actual exploration of AI-assisted forecasting revealed—what the proof of concept tested, what it found at each time horizon, and what it suggested about where this approach might add genuine value in a treasury context.

CHAPTER 6

Exploring AI-Assisted Cash Forecasting

What a Proof of Concept Revealed About Cash Forecast Horizons

This chapter describes what happened when the cash forecasting architecture was built and tested on real treasury data patterns. The proof of concept described here was designed and tested in mid-2026— readers should interpret the findings in that temporal context, recognizing that the underlying models and tooling are evolving continuously. The proof-of-concept results—including what worked well, what the findings revealed at each horizon, and what remains to be demonstrated at production scale—are presented honestly. It includes the dashboard built for treasury users, with the design decisions that made it useful and accessible.

Chapter 5 set out the problem: why cash forecasting is genuinely difficult, what good forecasting would actually look like, and why AI-based approaches are worth exploring. What follows is what happened when we moved from that framing to actual exploration — building a working proof-of-concept that applied a three-model ensemble to a synthetic treasury dataset modeled on real treasury transaction patterns, and observing what it produced at each time horizon.

The findings are presented here as what they are: the results of a controlled proof-of-concept exploration, not a validated production system. A POC can show that an approach is technically feasible and produces encouraging results on representative data. It cannot show that the approach will operate reliably across the full complexity of a real enterprise treasury environment. That distinction matters, and it will be maintained throughout this chapter. The three model families used—ARIMA, XGBoost, and Prophet—were selected as a representative subset suited to the forecasting horizons being tested. Many other tools and frameworks are available, and the choice of these three reflects the requirements of this particular exploration rather than any general preference.

The Dashboard: Seven Views of the Forecasting Problem

The proof of concept was built as a working Streamlit application with seven tabs, each serving a distinct purpose in the treasury workflow. **FIGURE 6.1** shows the dashboard navigation structure—the first thing a treasury user sees when opening the system.

All figures in this chapter are drawn from the proof of concept and show synthetic data only. The "Learn more about this view" sections built into each tab were written deliberately in non-technical language. The target user is a treasury business user, not a data scientist. This editorial choice reflects a core argument of this book: AI tools in treasury need to be accessible to treasury professionals, not just to technologists. If the system cannot explain itself in treasury language, it will not earn the trust it needs to be used.

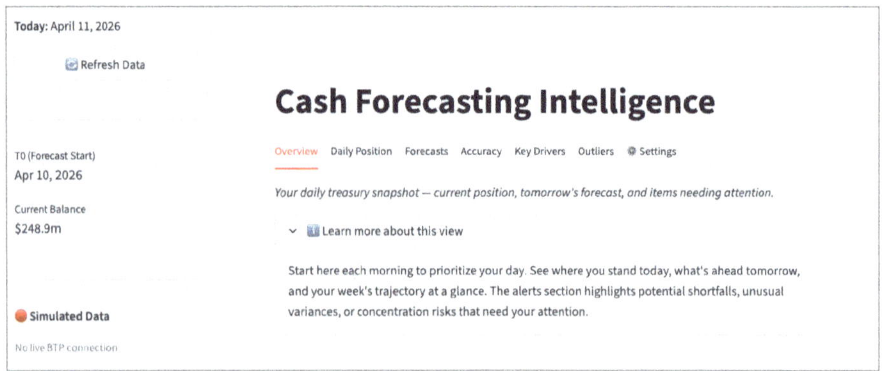

FIGURE 6.1: The proof-of-concept dashboard showing the seven-tab navigation structure: Overview, Daily Position, Forecasts, Accuracy, Key Drivers, Outliers, and Settings.

What We Set Out to Test

The Chapter 5 introduced the ensemble concept — combining multiple models weighted by horizon strength rather than relying on any single model across all horizons. The POC was built to test that concept in practice: does an ensemble of three models, weighted differently by horizon, produce meaningfully better forecasts than a single model applied across all horizons? The benchmark was a Prophet-only model applied to the aggregate closing balance — a reasonable baseline representing something more sophisticated than a spreadsheet but simpler than a full ensemble.

The synthetic dataset reflected the pattern range a treasury user at a mid-to-large corporation would typically encounter: accounts receivable collections with day-of-week effects and billing cycle seasonality; accounts payable disbursements with end-of-month concentration; biweekly payroll; monthly debt service; quarterly tax payments; and intercompany flows. Banking day awareness was built in.

The Morning View: Overview and Executive Summary

The first question any treasury user brings to a forecasting system is the simplest one: where is my cash position heading? The Overview tab was designed to answer that in thirty seconds. **FIGURE 6.2** shows the executive summary section, including the four metrics that matter most at the start of the day.

The confidence level indicator — HIGH, MEDIUM, or LOW — is not cosmetic. It reflects the model's internal uncertainty about the forecast given the data available that morning. A HIGH confidence forecast on a day with confirmed bank feeds and clean TMS data is different from a MEDIUM confidence forecast produced when intraday data has not yet arrived. The treasury user needs to know the difference.

In the POC, this is a recommendation—the human decides whether to act. That distinction is worth preserving in any production implementation.

Executive Cash Position Summary

Last updated: 2026-04-11 11:01:40 | T0: 2026-04-10

Today's Position — Monday, Apr 13 2026

Opening Balance	Projected Closing	Net Change	Confidence
$248.9m	$249.8m	$875.5k ↑ $875.5k	● HIGH

Recommendation: INVEST $239.8m—Excesscash$249,767,798. Consider overnight investment.

FIGURE 6.2: The POC executive cash position summary.

T+1: Daily Cash Positioning as a Living Document

At the T+1 horizon, the ensemble is dominated by ARIMA, with Prophet providing minor supporting weight. ARIMA's short-term autoregressive patterns — its ability to extrapolate from very recent history — produced meaningfully lower error at the one-day horizon than a Prophet-only model in the POC comparison.

At the T+1 horizon, the most important finding was not about aggregate accuracy but about the architecture of the position itself. Knowing that a projected shortfall is driven by the AR lockbox running below forecast calls for a different response than one driven by an AP disbursement or an intercompany timing issue. The category breakdown makes that distinction visible. The aggregate number cannot.

FIGURE 6.3 shows the category-level position detail that makes this distinction actionable.

The Status column is the key innovation here. It distinguishes between "Confirmed from bank" (the opening balance), "Posted (4 transactions)" (cleared transactions with count), "Estimated" (still forecast), and "SAP: $108,471" (scheduled payments sourced from the ERP system). This source traceability is what makes the position auditable and the morning conversation with the CFO or the banks productive.

Position Worksheet

Category	Forecast	Estimated Actual	Variance	Status
Opening Balance	$248.9m	$248.9m	$0	Confirmed from bank
AR - Wire Receipts	$556.0k	$444.8k	-$111.2k	Estimated
AR - ACH Credits	$556.0k	$1.1m	$543.4k	Posted (4 txns)
AR - Lockbox	$556.0k	$391.4k	-$164.6k	Posted (1 txns)
Investment Income	$0	$0	$0	-
Intercompany In	$184.4k	$147.6k	-$36.9k	Estimated
Other Receipts	$0	$0	$0	-
TOTAL RECEIPTS	$1.9m	$2.1m	$230.7k	
Payroll	$0	$0	$0	-
AP - Wires	$0	$108.5k	$108.5k	SAP: $108,471
AP - ACH Payments	$0	$970.2k	$970.2k	SAP: $970,183
AP - Check Clearings	$0	$128.9k	$128.9k	Cleared: $128,938
Tax Payments	$0	$0	$0	-

FIGURE 6.3: The POC daily Position Worksheet showing category-level receipt detail — AR Wire Receipts, AR ACH Credits, AR Lockbox, Investment Income, Intercompany In — with Forecast, Estimated Actual, Variance, and Status columns.

T+7: Where Feature Patterns Begin to Dominate

At the seven-day horizon, the forecasting problem changes character. Recent trend continuation becomes less informative. What matters at T+7 is feature interaction: day-of-week effects on AR collections, end-of-week AP settlement patterns, proximity to payroll dates, month-end concentration effects. This is where XGBoost's pattern recognition capability proved most valuable in the POC.

FIGURE 6.4 shows the T+7 view using a candlestick-style chart — a deliberate design choice for an audience of treasury professionals who are familiar with this format from FX and investment dashboards.

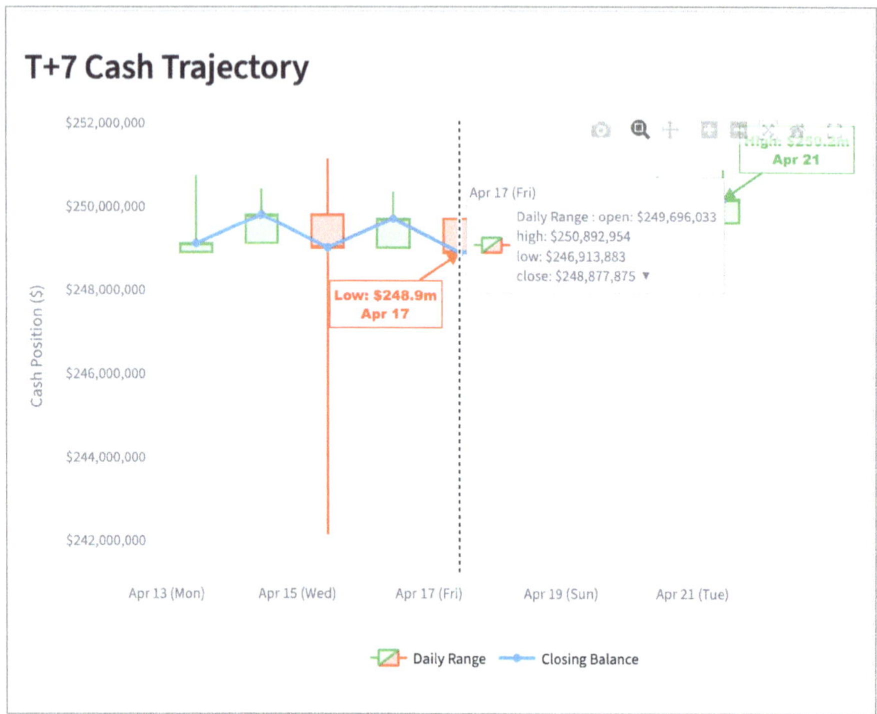

FIGURE 6.4: The POC T+7 forecast as a candlestick chart — green candles for net inflow days, red for net outflow days — with the closing balance line overlaid and min/max balance annotations.

The min and max balance annotations show the expected range for the week at a glance, helping the treasury user plan funding and investment decisions.

A particularly useful finding at T+7 was the day-of-week accuracy pattern: forecast accuracy was not uniform across the week. The Accuracy tab revealed that certain days showed more variability than others — reflecting settlement timing effects that are well-known to treasury practitioners but that the model needed to learn empirically. This kind of insight changes how the treasury user applies the forecast: higher scrutiny on less reliable days, more confidence on historically stable ones.

T+30: Structural Patterns and Working Capital Planning

At the thirty-day horizon, the forecasting problem becomes primarily structural. Short-term noise has largely averaged out. Monthly seasonality, known scheduled events, and the underlying business trend dominate the forecast. Prophet carries the most weight in the ensemble here, and the category-level forecasting approach shows its clearest advantage (see **FIGURE 6.5**).

The dashed vertical lines marking tax payments and debt service — registered in the Settings configuration as known future events — are a good illustration of the partnership between AI pattern learning and treasury professional domain knowledge. The model cannot predict a debt service payment it has not been told about. By registering known

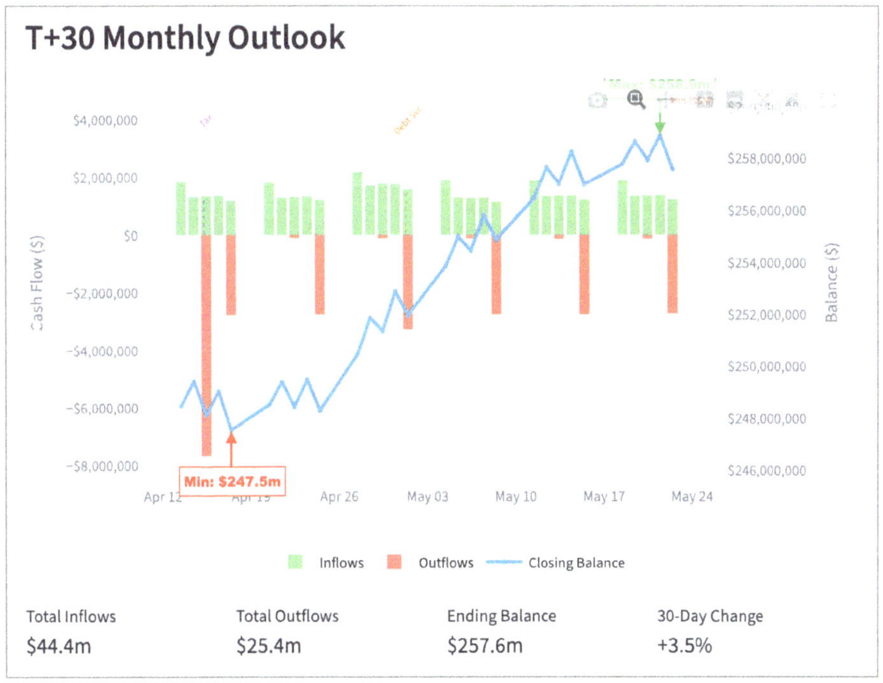

FIGURE 6.5: **The POC T+30 Monthly Outlook — daily inflows (green) and outflows (red), closing balance trajectory (blue), and dashed vertical lines marking known scheduled events (tax, debt service). Summary metrics show total inflows, outflows, ending balance, and 30-day change.**

future events, the treasury user ensures these are incorporated as fixed inputs rather than predicted from history – which meaningfully improves longer-horizon accuracy.

The mid-period dip visible in the trajectory chart is an early warning of a tight period – seen weeks in advance, when there is still time to act.

The T+30 view supports working capital planning decisions: should I consider early payment to capture a discount, or delay to preserve cash? Is there sufficient projected coverage for the debt service payment at day 22? The forecast does not make these decisions – but it provides the context in which a treasury professional can make them with more confidence.

The Accuracy Finding

The Accuracy tab provided the most direct evidence on the central question: does the ensemble outperform the Prophet-only baseline, and by how much? **FIGURE 6.6** shows the accuracy comparison across all four horizons.

FIGURE 6.6: POC accuracy metrics – balance error 0.56% (Excellent), inflow error 6.57%, outflow error 5.13% – with breakdown by horizon day (left) and day of week (right).

On a $250m balance, 0.56% error means the model was typically within $1.4m — well within the range that supports confident daily investment and funding decisions. The more useful findings, however, are in the details.

The Balance Accuracy by Horizon Day chart shows that error stays below 2% through day 6, then rises at day 7. This is actionable: T+5 may be the practical limit for high-confidence decisions, with days 6 and 7 treated as directional guidance rather than precise forecasts.

The Accuracy by Day of Week chart surfaces something more specific: Wednesday outflows show dramatically higher error than other days. In a real treasury environment, this would prompt an investigation: is there a specific weekly payment cycle (perhaps vendor payments processed on Wednesdays) that the model is not capturing well? This kind of finding is what the archive-and-accuracy feedback loop is designed to reveal — systematic patterns that can be investigated, explained, and corrected.

The Explainability Finding

One of the most important architectural decisions in the POC was the Key Drivers tab—the explainability layer that answers this question: why is the forecast what it is? **FIGURE 6.7** shows the factor importance analysis for inflows and outflows.

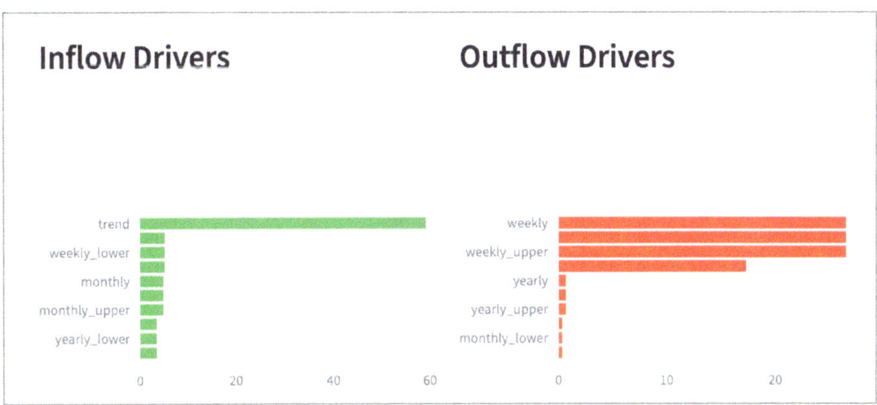

FIGURE 6.7: The POC Key Drivers screen showing factor importance for inflows (left, green) and outflows (right, red). Inflows are dominated by trend (~58), with weekly, monthly, and yearly seasonality contributing smaller adjustments. Outflows are driven primarily by weekly seasonality (~28), reflecting fixed-day payment run patterns.

The contrast between the two charts is structurally revealing: inflow patterns are dominated by business trend — the underlying growth trajectory of the business drives receivables — while outflow patterns are dominated by weekly seasonality, reflecting the reality that AP departments typically run payment cycles on fixed days.

For treasury, this matters because it validates the category-level forecasting approach. A model that treats inflows and outflows identically would miss these structural differences. Forecasting AR and AP separately, using Prophet's decomposition, captured this naturally.

The explainability layer is also what enables the override. When the treasury user can see that the forecast is driven primarily by "typical Tuesday collection patterns," they can assess whether those patterns apply this Tuesday — and apply a targeted override if they know something the model does not.

The Outlier Detection Finding

The Outliers screen (**FIGURE 6.8**) illustrates perhaps the clearest transition from passive prediction to active treasury management.

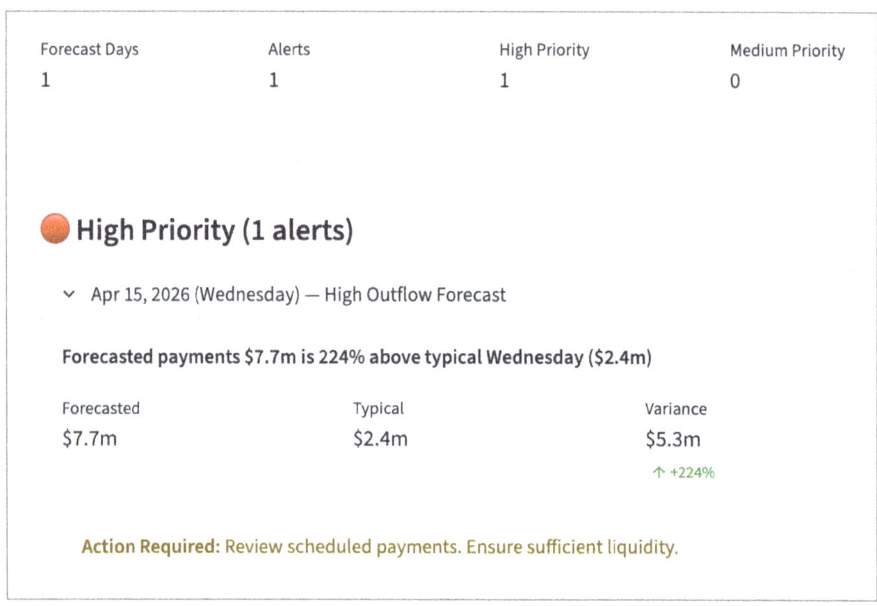

FIGURE 6.8: POC High Priority outlier alerts — each showing Forecasted, Typical, and Variance values alongside an Action Required recommendation.

Each alert surfaces a forward question: does the forecasted pattern for this day match what typically happens? When it does not, the system prompts a targeted check — in this case, verifying whether payments have been postponed or mis-scheduled, or are pending approval.

Not every outlier is a problem. Sometimes a deviation is expected — the treasury user knows a large customer is paying early, or a supplier invoice is being held. The value is in surfacing the question before the day arrives, when there is still time to investigate and act. A false alert costs thirty seconds of attention. A missed alert — a cash shortfall that was not seen coming — can cost overnight borrowing fees, missed investment returns, or worse.

What the Exploration Did Not Resolve

Honest assessment requires naming what the POC did not answer as clearly as what it did.

The POC used synthetic data. Whether the ensemble accuracy advantage holds when the training data contains the structural breaks, missing periods, and categorization changes that characterize real multinational treasury datasets is a genuine open question.

The POC was a single-entity, single-account model. How the architecture scales to dozens of entities, multiple currencies, and intercompany relationships requires production environment testing to answer.

The POC could not test organizational adoption. A technically robust forecasting system that the treasury team does not trust, does not use, and does not maintain will not deliver value. These are as important as the technical dimensions and cannot be tested in a controlled POC environment.

These are not reasons to dismiss the POC findings. They are reasons to approach any production implementation with the careful, incremental, data-quality-first approach described in the 90-day pilot framework in Chapter 11.

PUT IT INTO PRACTICE

Assess your forecasting data readiness for an AI-assisted approach.

- How many years of daily closing balance history do you have in a structured, accessible format? Two years is a minimum for reliable seasonal pattern learning. Consider all potential data sources: your TMS or ERP system for posted cash transactions, bank statement feeds (MT940, BAI2, or camt.053), intercompany settlement records, and any subsidiary forecast submissions captured systematically.
- Can you extract category-level history — AR, AP, payroll, intercompany — separately and consistently across that history?
- What are your major scheduled cash flows over the next 90 days? In an AI-assisted system these would be registered as known events, meaningfully improving longer-horizon accuracy.
- Do you have a single account or entity where the data is cleanest? That is your starting point for any pilot.
- What does your current T+7 and T+30 forecast accuracy look like? If you do not know, establishing that baseline is the first step.

CHAPTER SUMMARY

- The proof-of-concept tested a three-model ensemble against a Prophet-only baseline on synthetic data modeled on real treasury patterns; it found consistent accuracy improvement at all horizons.
- These findings are presented as what they are: results from controlled exploration on representative synthetic data, not evidence of production-validated performance at enterprise scale.
- Each time horizon serves a different treasury decision: T+1 for same-day funding and investment; T+7 for short-term liquidity management; T+30 for working capital planning; T+90 for directional strategic guidance.
- Category-level forecasting—AR, AP, and payroll modeled separately— proved more useful than aggregate balance forecasting because it reveals the composition of any variance and points to the appropriate treasury response.
- The Key Drivers explainability layer enables the override: when the treasury user can see why the model is predicting what it is predicting, they can assess whether those reasons apply and act on information the model does not have.
- Outlier detection converts a passive forecasting tool into an active decision-support system; the value is in seeing potential problems before the day they arrive.
- The POC also evaluated the T+90 horizon. At that range, structural patterns dominated and the ensemble produced directional guidance rather than actionable precision—consistent with the horizon limitations described in Chapter 5. A T+90 view is most useful as a strategic planning input rather than an operational decision tool.
- Three important open questions remain: how accuracy holds on real enterprise data; how the architecture scales to multi-entity complexity; and how organizational adoption develops in practice.

In the next chapter, we turn to intelligent hedge management. This is an area where the complexity of the problem, the stakes of getting it wrong, and the potential for well-designed automation to reduce manual burden make it equally worth the same careful, deliberate approach that cash forecasting demands—and in some respects, the challenges are distinctly different in character.

CHAPTER 7

Intelligent Hedge Management

From Fragmented Process to Orchestrated Intelligence

Hedge management involves more interdependent components than almost any other treasury process—exposure identification, policy application, intercompany netting, trade execution, accounting documentation, and reconciliation, all connected and time-sensitive. This chapter maps that process end-to-end, and explains where a well-designed multi-agent architecture adds the most value.

Of all the areas in corporate treasury where I have spent time over the years, hedge management is the one that has most consistently illustrated the distance between what the process design looks like on paper and the operational reality of running it in a complex global organization. Not because the underlying concepts are especially opaque—they are not—but because the end-to-end hedge management process in a large multinational organization involves so many interdependent components, so many system handoffs, and so many points where human judgment must be applied under time pressure, that even well-resourced treasury teams find it demanding to keep it running smoothly.

Hedge management is consistently one of the highest-friction areas of the treasury function. It is also one of the areas where agentic AI has the greatest potential to deliver genuine, lasting improvement—not by replacing the judgment that treasury professionals bring to it, but by handling the orchestration burden that currently consumes so much of their time.

This chapter walks through the entire hedge management process as I have observed it operating in complex organizations—where the work is concentrated and where the friction accumulates—and how a well-designed multi-agent architecture can begin to change the picture in meaningful ways. The examples that follow draw on SAP treasury implementations, but the process challenges and the principles behind the solutions are universal; they apply regardless of which treasury management system an organization uses.

The Full Arc of Hedge Management

It is worth mapping the complete process before discussing where AI fits, because the complexity is not obvious until you lay it out in its entirety. What follows is that map, drawn from the design and build process currently underway.

Identifying Raw Exposures

The process begins with exposure identification. Before any hedge decision can be made, treasury needs to understand what the organization is actually exposed to. In a single-entity business, this is relatively contained. In a global corporation with subsidiaries operating across multiple currencies, legal jurisdictions, and booking systems, it is anything but.

The first distinction that matters—and one that is worth maintaining carefully throughout the process—is between cash flow exposures and balance sheet exposures. They are different in nature, they carry different accounting implications, and they require different hedging approaches.

Cash flow exposures arise from anticipated future transactions: a purchase order denominated in euros, a contracted customer receipt in Japanese yen, a forecasted intercompany payment from a subsidiary in a restricted currency market. These exposures exist before any accounting entry has been made. Hedging them effectively requires a degree of confidence in the underlying forecast, and if the organization wishes to apply hedge accounting under International Financial Reporting Standard 9 (IFRS 9), the international accounting standard governing financial instruments, including hedge accounting, or Accounting Standards Codification 815 (ASC 815), the US Generally Accepted Accounting Principles (GAAP) standard governing derivatives and hedging, it must formally designate each hedging relationship, document the linkage between hedge and hedged item, and demonstrate ongoing hedge effectiveness. The documentation burden alone is considerable.

Balance sheet exposures arise from items already recorded—foreign currency receivables, payables, and intercompany loans. These are real, current obligations. They do not require the same forecast confidence as cash flow hedges, but they introduce their own complexity: the exposure changes every time a new transaction posts, which means the hedge position must be reassessed at each measurement date, and any mismatch between the hedge and underlying exposure must flow directly through the income statement.

Getting this classification right at the point of identification—and keeping it current as business activity continues—is one of the areas where the process most benefits from systematic support. A significant part of the problem is structural: exposure data rarely lives in a single system. Receivables sit in one module, payables in another, intercompany balances in a third, and treasury deals in the TMS. Pulling these together into a coherent, current exposure view requires either manual effort or well designed integration—and in many organizations, it is the former. The exposure register is not inaccurate exactly; it is simply not current. By the time someone has extracted the data, reviewed it, classified each item, and prepared a hedge recommendation, the underlying exposures may have already changed.

Policy Constraints: Hedge Ratios, Accounting Treatment, and Documentation

Layered on top of the exposure identification challenge are the organization's own internal policies, which introduce further parameters that every step of the process needs to reflect.

Most treasury policies specify a target hedge ratio—the proportion of identified exposure that should be hedged—typically expressed as a range rather than a fixed number. A policy requiring hedging of between 50 and 90 percent of forecasted foreign currency cash flows gives treasury room to exercise judgment, but it also means that every hedge decision requires a check of the current hedge position against the permitted range, followed by a determination of whether any action is required. When managing exposures across twenty or thirty currency pairs and dozens of legal entities, this is not a trivial exercise.

The choice of accounting treatment introduces another layer of complexity. Shortcut-method hedge accounting, available under certain conditions in ASC 815, assumes perfect effectiveness and avoids the requirement for ongoing quantitative effectiveness testing. It is administratively simpler, but its eligibility criteria are narrow. Full fair value or cash flow hedge accounting is more flexible in application but demands rigorous documentation: formal hedge designation at inception, description of the risk management objective, identification of the hedged item and hedging instrument, the method to be used for assessing effectiveness, and ongoing measurement at each reporting date.

Organizations frequently maintain both shortcut and full hedge accounting relationships simultaneously, with different treatments applied to different exposure types. Managing the documentation requirements across both—manually, in spreadsheets, against a reporting deadline—is exactly the kind of time-pressured administrative work where systematic support adds genuine value and where the precision AI can bring is most welcome.

The Intercompany Dimension

For global corporations, the hedge management process does not operate in each entity in isolation. It operates across a web of intercompany relationships—and this is where the true complexity of the problem becomes apparent.

Multilateral Netting: The Theory and the Reality

Intercompany netting is, in principle, an elegant mechanism. Rather than having each subsidiary settle its intercompany obligations individually—generating a large volume of individual cross-border payments, each with its own FX conversion cost and settlement risk—the organization calculates net positions across all entities and settles only the differences. The FX exposure of the group as a whole is reduced, transaction costs fall, and the treasury function gains a cleaner view of the group's true currency position.

SAP's In-House Banking module provides a robust foundation for implementing this model. In theory, the in-house bank sits at the center of the intercompany payment architecture: subsidiaries settle through the IHB rather than directly with external counterparties, and the IHB manages the consolidated FX position on behalf of the group.

In practice, the multilateral netting process across a global entity structure is an operationally demanding workflow—but one that, when implemented well, delivers results that justify the effort considerably. Several specific sources of friction recur consistently across organizations, and they are exactly the problem types that agent-based automation is well-suited to address.

Different Netting Rules Across Subsidiaries

Not all subsidiaries operate under the same netting arrangements. Some may participate in full multilateral netting; others may be on bilateral netting only, settling intercompany balances pairwise rather than through a central pool. Some subsidiaries may be excluded from netting entirely, either because local regulations require direct settlement or because the legal entity structure creates complications that make pooling inadvisable.

The netting rule structure across a large global organization tends to have evolved over time rather than been designed from scratch. Acquisitions bring entities with legacy settlement arrangements. Regulatory changes in specific markets force adjustments to what was previously a clean rule set. The result is a netting matrix that is rarely as uniform as the documentation suggests, and that requires careful handling at each netting cycle to ensure each entity is processed according to its specific rules rather than a common default.

Many organizations outsource the netting process entirely, to a specialist external system or a financial institution that runs the netting cycle on their behalf and provides settlement instructions based on the results. This approach can simplify the operational burden for the treasury team, but it introduces a different set of challenges: integrating the external netting output back into the primary book of record or TMS, reconciling positions across two systems that may use different entity identifiers or currency conventions, and maintaining a clean audit trail that ties the settled positions back to the original intercompany transactions. These integration and reconciliation challenges are, in my observation, frequently underestimated at the point of implementation.

None of this should be read as a reason not to pursue netting automation. Quite the opposite. Organizations that have implemented multilateral netting well—with clear entity rules, clean data flows, and a well-designed settlement process—consistently report results that justify the effort. Month-end intercompany close processes that once consumed weeks have been reduced to days. Transaction costs have fallen as external FX settlements are replaced by internal book entries. Treasury visibility into the group's true currency position improves significantly when the netting cycle runs reliably and on schedule. The complexity is real, but so is the payoff.

A further complexity is worth acknowledging: in many organizations, the intercompany netting process is not owned by treasury at all. It sits with a dedicated intercompany team that operates at the intersection of accounts payable and accounts receivable, with treasury involved primarily at the point of FX trade execution. This organizational separation can create gaps in data ownership and process coordination—treasury may not have visibility into the intercompany positions until the netting cycle is complete, by which point the window for timely hedge execution has sometimes already narrowed.

Country-Specific Restrictions and Cashless Versus Cash Settlement

Certain markets impose regulatory restrictions that affect how FX exposures can be managed. Some currencies are non-deliverable—meaning that FX risk can be hedged through non-deliverable forwards

(NDFs), FX hedging instruments used for currencies where physical delivery is restricted by local regulation, but physical settlement in that currency is not possible. Some jurisdictions require advance registration of hedging transactions with local regulators. Others impose restrictions on intercompany loans or cross-border cash movements that affect how netting settlements can be structured.

The distinction between cashless and cash settlement adds further complexity. Where netting results in a net payable from one entity to another, the organization may settle in cash—an actual transfer of funds between accounts—or through a book entry that records the intercompany obligation without moving cash. The choice between these approaches has tax implications, regulatory implications, and implications for how the resulting FX exposure is classified and measured. Different subsidiaries, in different jurisdictions, may need to be handled differently.

Managing these distinctions manually—entity by entity, currency by currency, settlement cycle by settlement cycle—is precisely the kind of rule-intensive, high-volume, low-tolerance-for-error work that occupies a disproportionate amount of treasury time in organizations without automation.

The Gap Between System Capability and Real-World Implementation

This is a point worth making carefully, because it is sometimes misunderstood.

SAP provides genuinely powerful tools for treasury management. The treasury and risk management (TRM) module, the In-House Banking functionality, the intercompany netting capabilities, the hedge accounting configuration options—including the Hedge Management Cockpit for cash flow hedges, trading platform integration, and direct integration with the general ledger for accounting entries—these represent a serious, mature platform that is more capable than most organizations fully utilize. The same is true of other leading treasury management systems. The functionality exists; the challenge is in realizing it consistently across the full process.

At the same time, even organizations running a well-configured TMS find that the end-to-end hedge management process requires substantial manual effort to bridge the gaps between its component parts. The issue is not that any individual module lacks functionality. It is that the hedge management process spans multiple modules—TRM, cash management, in-house banking, the general ledger, the payment program—and coordinating activity across those modules, in the right sequence, with the right data flowing between them, has historically required significant human effort at each handoff.

The following points in the process consistently generate manual effort, as shown in **TABLE 7.1**.

Each of these steps is manageable on its own. The opportunity that agentic AI addresses is that they are all interconnected, they all need to happen in a defined sequence, and they all need to happen with sufficient frequency to keep the hedge position aligned with the underlying exposure. Orchestrating that sequence reliably, at scale, is precisely what the multi-agent architecture is designed to do.

TABLE 7.1: When manual effort is required

Process Step	Typical Manual Effort Required
Extracting raw exposures	Pulling reports from multiple modules and entities, reconciling to a single exposure view
Classifying cash flow versus balance sheet	Reviewing each exposure line, applying policy criteria, resolving ambiguous cases
Applying hedge ratios	Checking current hedge position, calculating required trades, flagging policy breaches
Intercompany netting calculation	Running netting cycles, applying entity-specific rules, handling exceptions and outsourced netting reconciliation
Preparing trades for execution	Compiling trade details, obtaining approvals, formatting for bank or platform submission
Hedge accounting documentation	Preparing designation memos, effectiveness assessments, journal entry support
Reconciling positions	Matching executed trades back to the exposure register, updating positions, flagging discrepancies

Where Agentic AI Changes the Picture

The hedge management process, viewed through the lens of the agent architecture described in Chapter 4, is a natural fit for orchestrated sub-agent design. The process has a clear structure, involves well-defined rules (even where those rules vary by entity), requires coordination across multiple data sources and systems, and has a genuine need for human review at specific decision points—without requiring human involvement at every step.

Each major process step maps to a dedicated sub-agent, with an orchestrator coordinating their sequencing and managing the handoffs between them. In this example, there are five sub-agents, but in practical implementation, there are likely to be quite a few more (see Chapter 11, "What an Agent Pipeline Might Look Like.")

The Exposure Identification Sub-Agent

An exposure identification sub-agent would run on a defined schedule—daily, or more frequently for high-volume organizations—pulling FX-denominated open items from the TMS, accounts receivable, accounts payable, and the intercompany reconciliation module. It would apply the classification logic that distinguishes cash flow from balance sheet exposures, flag items that fall in ambiguous categories for human review, and produce a current exposure register without manual data extraction.

This sub-agent would also be responsible for maintaining the exposure register's currency as new transactions post. Rather than a point-in-time snapshot that ages from the moment it is produced, the exposure register becomes a living document, updated continuously, with a reliable audit trail of what changed and when.

The Netting Calculation Sub-Agent

A netting calculation sub-agent would handle the multilateral intercompany netting cycle, applying the correct netting rules for each participating entity, calculating net positions by currency pair, and identifying the settlement instructions required. This is exactly the kind of rule-intensive, entity-specific calculation that is well-suited to sub-agent execution—the rules are definable, the data inputs are structured, and the required output is precise.

Critically, this sub-agent would maintain the entity-specific rule set—which subsidiaries participate in multilateral netting, which are on bilateral arrangements, which require cash settlement versus book entry, which involve currencies with regulatory restrictions—as a configuration layer that can be updated as the organization's structure evolves, rather than embedded in manual process knowledge that walks out the door when a team member leaves. Where netting is outsourced to an external provider, this sub-agent can also manage the reconciliation of the external netting output back to the TMS, closing one of the most persistent integration gaps in the current process.

The Hedge Recommendation Sub-Agent

A hedge recommendation sub-agent would take the current exposure register, apply the organization's hedge ratio policy, check the existing hedge position, and calculate the trades required to bring the hedged proportion within the policy range. It would surface these as recommendations for treasury review—not as executed transactions—with the supporting rationale: current exposure, current hedge position, policy target, and the specific trades proposed.

For straightforward cases that fall clearly within policy parameters, this review step can be light. For cases at the boundary of policy limits, or involving currencies with restricted markets, or where the underlying exposure forecast has a higher degree of uncertainty, the sub-agent flags these specifically for deeper human review before any trade is prepared.

The Trade Preparation Sub-Agent

Once hedge recommendations have been reviewed and approved, a trade preparation sub-agent handles the mechanical work: confirming counterparty limits, formatting trade instructions for submission to the bank or trading platform, and maintaining the audit trail that links each executed trade back to the exposure it was designed to hedge—a linkage that is essential for hedge accounting documentation.

The Reconciliation Sub-Agent

After trades are executed and settlements occur, a reconciliation sub-agent matches executed trades back to the exposure register, identifies any discrepancies between the expected and actual settlements, and updates the hedge position accordingly. Where positions are out of alignment—perhaps because an underlying exposure has changed since the hedge was executed—the sub-agent flags the mismatch and calculates the rebalancing action required.

The Orchestrator

Sitting above all of these is the orchestrator, which coordinates the sequence of sub-agent activity, manages the timing of each step relative to the overall hedge management cycle, escalates exceptions that require human decision-making, and maintains the overall view of where the process stands at any point in time. The treasury professional's interaction with the system is primarily at the orchestrator level—reviewing the current hedge position, acting on flagged exceptions, approving recommendations before they proceed to execution, and monitoring the overall health of the hedge program. Chapter 11 goes into more detail about how to construct the orchestrator, in the section "The Orchestrator as Policy Document."

This is the vision of agentic AI in hedge management: not a system that makes hedge decisions autonomously, but one that handles the orchestration burden that currently consumes treasury capacity—freeing the professionals who understand the organization's risk strategy, its accounting obligations, and its regulatory environment to focus on the judgment calls that genuinely require their expertise.

What the Architecture Demonstrates

The FX exposure management workflow described in this chapter is not just a conceptual illustration—the architecture is being built and tested as a proof of concept using real treasury data patterns. What it demonstrates is worth reflecting on directly, because the lessons apply beyond FX management to the broader question of how AI should be designed for enterprise treasury.

Functional Expertise Is the Core Asset

Every policy rule in the orchestrator, every field name in the data extraction specification, every netting logic decision, every IFRS 9 designation criterion—these came from functional treasury knowledge, not from a developer. The AI executes. The treasury expert designs. This is the right division of labor, and it is the principle that makes these systems both trustworthy and maintainable. A system built by technologists who do not understand treasury may encode the wrong assumptions. A system designed by treasury professionals who understand what the rules are, why they exist, and where the edge cases live will encode the right ones.

Single Responsibility Makes Systems Safe

A single AI system that attempts to extract data, net positions, size hedges, check compliance, and post deals in one unstructured flow is difficult to audit and difficult to diagnose when something goes wrong. The single-responsibility agent architecture makes failures diagnosable: when the netting engine produces an unexpected result, the issue is isolated to that sub-agent, its inputs, and its rule set. Outputs are auditable at every stage. Policy changes are surgical rather than systemic. This is not an abstract architectural preference—it is what makes the difference between a system an auditor can evaluate and one that cannot be meaningfully reviewed.

The Guardrail System Is Treasury Policy Made Executable

Every check in the orchestrator's validation layer corresponds to a treasury policy rule that would exist in a policy document. The innovation is not the rule itself. It is the fact that the rule now actively controls whether the pipeline runs, rather than being applied retrospectively by an analyst reviewing outputs after the fact. Policy enforcement moves from periodic audit to real-time gate. This is a meaningful shift in how treasury controls operate in practice, and one that treasury professionals who have spent time chasing policy exceptions after the fact will recognize immediately.

Enterprise Resource Planning Integration Is Achievable Without Disruption

A concern that frequently arises when treasury technology teams consider AI integration is the risk of disrupting existing ERP environments. The architecture described in this chapter addresses that concern directly. Using SAP as an example, the system connects through Business Technology Platform (BTP) —the standard, approved, supported integration layer—with no custom development, no direct database access, and no bypass of SAP's security model. IT and information security teams can approve this architecture because it uses the access patterns the ERP system already sanctions. The same principle applies to other leading ERP and TMS platforms: the integration approach should work with the system's standard API and connectivity layer, not around it.

A Practical Observation on Implementation: Data and Process First

It is easy to describe the target state in a way that undersells the work required to get there.

The sub-agent architecture described above can only function well if the underlying data is reliable. An exposure identification sub-agent pulling from a system where intercompany accounts are inconsistently maintained, or where the cash flow versus balance sheet classification has never been applied rigorously, will produce an exposure register that reflects those underlying problems. Automation amplifies what is already there—it does not correct structural data quality issues.

Similarly, the netting calculation sub-agent is only as good as the rule set it is working from. If the entity-specific netting rules have never been fully documented—if they exist primarily as institutional knowledge held by one or two people in the treasury team—the first task is to surface and codify those rules before any automation is built on top of them. This is often valuable work in its own right, independent of the AI implementation. It is also worth noting that in many organizations, this codification exercise requires active collaboration with the intercompany team in AP and AR, who often hold parts of the process knowledge that treasury does not.

Organizations that approach agentic AI in hedge management as a data and process discipline first, and a technology implementation second, will be better positioned to achieve durable results than those who lead with the technology and discover the issues later. The 90-day pilot framework described in Chapter 11 offers one structured way to sequence this work. And the organizations that do this well find that the effort is genuinely transformative—not just in the efficiency of the hedge management process itself, but in the quality of the risk picture treasury can provide to the CFO and the board.

PUT IT INTO PRACTICE

- Map the manual handoffs in your current hedge management process. Identify the five or six points where a human being is currently required to extract data, apply a rule, or prepare an output. These are your highest-priority automation candidates.
- Review your exposure classification practice. Is the cash flow versus balance sheet distinction being applied consistently and currently, or is it a periodic exercise done at month-end? The answer will tell you a great deal about the reliability of your current exposure register.
- Document your intercompany netting rules by entity. If the rules that govern which subsidiaries participate in multilateral netting, and on what terms, exist primarily as tribal knowledge, codifying them is a valuable first step regardless of whether you pursue AI implementation.

- Assess your hedge accounting documentation process. If designation memos, effectiveness assessments, and the supporting linkage between hedge and hedged item are being produced manually against a reporting deadline, this is a high-risk area that automation can address—but only if the underlying trade-to-exposure linkage is being maintained cleanly in your TMS from the point of execution.
- If your netting process is outsourced to an external provider, map the current reconciliation workflow between that provider's output and your primary TMS. This integration point is one of the most tractable automation opportunities in the hedge management cycle.
- Identify one part of the hedge management cycle where the sub-agent concept could be piloted. The netting calculation is often a good starting point: the rules are definable, the data inputs are structured, and the value of automation is immediately visible to the team.

CHAPTER SUMMARY

- Hedge management in a global organization involves a connected chain of activities—exposure identification, policy application, netting, trade execution, accounting documentation, and reconciliation—that is rarely as smooth in practice as it appears on paper.
- The distinction between cash flow and balance sheet exposures is fundamental and worth maintaining carefully throughout the process; misclassification has flow-on effects through hedge accounting treatment, documentation requirements, and income statement impact.
- Internal policies governing hedge ratios and accounting treatment (shortcut versus full) add further parameters that need to be reflected at every step of the process.
- For global corporations, the intercompany dimension—multilateral netting, entity-specific settlement rules, country restrictions, and cashless versus cash settlement—adds significant operational complexity that is a consistent source of manual effort.
- Many organizations outsource netting to external providers, which introduces integration and reconciliation challenges back to the TMS that are frequently underestimated at implementation.
- The outsourced netting integration point—reconciling the external provider's output back to the primary TMS—is one of the most tractable and highest-value automation opportunities in the hedge management cycle; a netting calculation sub-agent can close this gap systematically rather than through periodic manual reconciliation.

- As an example, SAP provides a capable foundation for treasury management—including the hedge management cockpit, trading platform integration, and accounting connectivity—but the process spans multiple modules, and coordinating activity across those modules has historically required substantial human effort at each handoff. The same gap exists across other leading treasury management systems.
- A multi-agent architecture—with dedicated sub-agents for exposure identification, netting calculation, hedge recommendation, trade preparation, and reconciliation, coordinated by an orchestrator—maps naturally onto the structure of the hedge management process.
- Successful implementation requires data quality and process discipline as prerequisites, and often requires collaboration with intercompany, AP, and AR teams whose process knowledge treasury does not always hold. Organizations that lead with process discipline achieve more durable results.

In the chapters that follow, we turn from the treasury workflows that agentic AI can support to the foundations that make any AI implementation trustworthy.

Chapter 8 addresses the data architecture that feeds it; Part III covers the explainability practices that make its outputs defensible, and—critically—the guardrails and governance frameworks that keep it operating within appropriate boundaries. Of everything this book covers, the governance and control layer deserves particular attention in hedge management: the combination of market risk, accounting obligation, and regulatory exposure means that the cost of an uncontrolled error here is higher than in almost any other part of the treasury function. Each of these foundations is as important to get right as the technology itself.

CHAPTER 8

The Data Foundation

Why the Quality of Your Treasury Data Determines the Quality of Your AI

Good data is the foundation of every major treasury transformation—ERP implementations, TMS deployments, and AI initiatives alike. This chapter explains where treasury data lives, how to assess your organization's specific readiness, and—importantly—how AI is not just a consumer of good data but a powerful tool for improving it through cleansing, harmonization, error detection, and ongoing quality management.

Every chapter in this book has touched on data in some form. The forecasting engine in Chapter 5 is only as accurate as the transaction history it learns from. The agent architecture in Chapter 4 depends on clean, structured inputs flowing between agents. The fraud detection systems in Chapter 3 learn their patterns from historical payment data. The daily cash positioning system in Chapter 6 is only as current as the intraday feeds that power it.

Data is not a supporting character in the AI treasury story. It is the foundation on which everything else is built. This is not unique to AI—every major treasury transformation, from an ERP implementation to a TMS deployment to a reporting overhaul, depends on the same underlying data quality. The organizations that have invested in clean, consistent, well-documented data over the years find that investment pays dividends across every system change, and AI is no exception.

What is distinctive about AI is that it can work with you on data quality at two distinct stages. During the build phase, AI can actively assist with data cleansing and harmonization—identifying duplicate records, flagging inconsistent classifications, suggesting mappings between local and global chart of accounts structures, and surfacing gaps in historical coverage that would otherwise only be discovered mid-project. This is preparation work that traditionally requires significant manual effort, and AI can accelerate and sharpen it considerably. In ongoing operations, the same capabilities continue to add value: anomaly detection surfaces unusual patterns in incoming data, pattern recognition identifies when a feed is behaving differently from its historical norm, and consistency checks flag when a categorization has drifted from its established definition.

Error detection is part of the same picture. A large, structured report—fifty rows, ten columns, hundreds of date entries—is exactly the kind of dataset where a single typo is easy to miss in manual review, but trivial for an AI validation check to catch. A date entered as 2005 instead of 2025 does not match the expected pattern; it gets flagged immediately. The value is not that the error was sophisticated—it is that AI validation is systematic and tireless in a way that human review of large datasets cannot reliably be. The relationship between AI and data quality is not simply one-directional. Used thoughtfully, AI is as much a tool for understanding and continuously improving your data as it is a consumer of it.

This chapter addresses that challenge directly—the practical reality of what treasury data actually looks like, where it lives, what it takes to make it AI-ready, and how to approach that work systematically without consuming your entire implementation timeline. It is worth noting at the outset that many organizations are already well-positioned. A treasury team that has maintained a TMS or ERP system consistently, reconciles positions regularly, and has a clean record of banking data is already in good shape. The challenges described in this chapter are real—but they are not universal, and the data readiness assessment later in this chapter is what tells you where you actually stand.

Understanding Your Treasury Data

Treasury data presents a specific set of challenges that are worth understanding clearly before any AI implementation begins. These are not unique to any particular organization—they reflect the structural realities of how treasury systems have been built and evolved over time.

Treasury data lives in multiple systems that don't speak the same language. A multinational treasury operation typically draws data from an ERP system or TMS for accounting entries, cash positions, and deal records; multiple banking platforms delivering balance and transaction data in BAI2, MT940, or camt.053 formats; FX trading platforms for market rates and executed trades; intercompany systems for netting positions; and sometimes separate payroll, AP, and AR systems that feed into the treasury picture but are owned by other departments. Each system was designed independently, uses its own data model, and represents the same underlying reality—a cash movement, a currency exposure, a payment—in its own way.

Treasury data carries the history of organizational change. Every chart of accounts reorganization, every system upgrade, every acquisition, every change in banking relationships leaves a mark on the data. Categories that meant one thing in 2018 may mean something slightly different today. A subsidiary acquired in 2020 may have brought its own data conventions that were not fully harmonized. A machine learning model trained on this history will learn those patterns along with the genuine signal. If a migration caused a period of changed categorization, the model will treat that period as a real pattern unless someone explicitly identifies and handles it.

Treasury data has a business context that the data itself doesn't capture. The single largest payment in a transaction history might be a one-time debt refinancing that will never recur. An unusual cash pattern in Q3 2020 was driven by pandemic-related supply chain disruption. A spike in intercompany activity in December may reflect a year-end netting cycle rather than a structural change in the business. A model without access to this context will either learn these as patterns to extrapolate—potentially producing misleading forecasts—or treat them as noise to filter, which risks removing genuine signal along with the anomalies.

Where Treasury Data Lives

Understanding the landscape of data sources available for AI implementation is the starting point for any readiness assessment. Treasury data comes from a broader set of systems than is often initially recognized.

The Book of Prime Record

The *book of prime record*—the system in which treasury transactions are initiated, recorded, confirmed, and settled—is typically the richest and most authoritative source of treasury data. In most multinational organizations, this is an ERP system such as SAP, Oracle, or a dedicated treasury management system such as Kyriba, ION, or FIS. The specific platform matters less than the discipline with which it has been maintained.

From this source, AI applications can draw on cash and liquidity data (posted accounting entries, end-of-day and intraday positions), payment data (payment runs, approval chains, timing, method, counterparty), deal records and exposure calculations for FX and interest rate risk, and intercompany balances and netting positions. The reliability of this data for AI purposes reflects the configuration and maintenance discipline applied over time. Systems that have maintained consistent chart of accounts structures, entity hierarchies, and document classification will generally yield more AI-ready data than those where these have evolved less consistently—not because of any fault in the system, but because of how enterprise systems naturally evolve in complex organizations over many years.

Banking Data

Banking data arrives through several channels, each with its own characteristics.

Prior day statements in BAI2 (North American standard) or MT940/camt.053 (ISO 20022) provide end-of-day balance and transaction details, which are posted to and update the book of prime record. This data is highly structured and well-suited for automated processing. One practical challenge is normalization—transaction type codes vary by bank, and the same transaction type may be coded differently across banking relationships.

Intraday statements in BAI2 or camt.052 format provide real-time updates throughout the business day. This data is essential for daily cash (T+1) positioning but requires more sophisticated processing—multiple updates per day, cumulative versus delta reporting conventions, and timing gaps between when transactions occur and when they appear in the statement. In most implementations, intraday data updates the cash position view but does not post to the general ledger.

SWIFT messages (MT series for traditional banking, ISO 20022 XML for more recent implementations) provide payment confirmations, bank confirmations for treasury deals, counterparty matching, and settlement notifications. For AI applications in trade confirmation and position reconciliation, this data provides authoritative external confirmation of treasury transactions.

Before building any AI application that depends on bank data, it is worth mapping transaction type code coverage across all banking relationships. The same economic transaction—an ACH receipt, a wire payment, a check clearing—may arrive with different type codes from different banks. Normalizing this mapping is foundational work that significantly improves dataset quality.

Planning and Forecasting Systems

A significant share of the forward-looking data that AI forecasting applications require does not come from transactional systems at all—it comes from planning and budgeting tools. Many organizations maintain cash flow forecasts, budget assumptions, and scenario models in dedicated planning platforms such as Anaplan, SAP Integrated Business Planning (IBP), or Oracle EPM. Others maintain this data in Excel workbooks—sometimes highly sophisticated ones representing years of refinement. In practice, all three approaches coexist in many large organizations, with different regions or business units using different tools and maintaining their own planning models.

Integrating planning data with AI forecasting models is one of the more consistently underestimated challenges in implementation. The structural issue is not the technology—it is how planning itself tends to work in practice.

In global corporations with decentralized planning structures, forecast data arrives from multiple regional sources, built on different assumptions, at different levels of granularity, and at different points in the planning cycle. The result is that by the time a consolidated view reaches treasury, it may already reflect varying degrees of currency across different parts of the organization.

Two patterns recur consistently in my experience. The first is that actual-versus-forecast reporting is often not maintained in a systematic, updated way—so the historical record of how planning forecasts compared to what actually occurred is incomplete or inconsistent. This matters for AI implementation because forecast accuracy history is exactly the kind of training signal that improves model performance over time. The second pattern is that cash flow forecasts in planning systems tend to be anchored to prior-year actuals rather than built from a fresh analysis of current trends. This is understandable—it is the fastest way to produce a defensible number—but it means the planning data carries a structural lag that an AI model needs to account for rather than learn from uncritically. Understanding the provenance and limitations of planning data is as important as understanding its content. An AI model trained on planning forecasts that were themselves anchored to last year's actuals will partially replicate that backward-looking tendency rather than correcting for it.

Legacy Systems

Many large organizations carry treasury-relevant data in legacy systems—older platforms for recording accounting entries, managing accounts payable, or processing accounts receivable that were never fully replaced when the organization modernized its core financial systems—or where that modernization is still underway. Systems are typically in the process of being replaced or consolidated, but that process rarely happens quickly. The interim period—which in practice can last years—means that historically rich data sits in systems that were never designed to integrate with modern AI pipelines.

The data quality challenges from legacy systems are specific and worth understanding before any AI project begins. Migrations are typically phased rather than a big bang-style cut-over, which creates its own complications. During the parallel-run period—when both the legacy system and the new system are running simultaneously—the same transaction may be recorded in two systems with different coding conventions, different timing, and potentially different amounts due to currency or rounding treatment. Any AI model drawing training data from a period that spans a phased migration will encounter this structural break and require explicit handling to exclude or adjust the affected period. Mergers and acquisitions add a further layer of complexity. The practical question for AI readiness is not whether to use legacy data—the historical depth it provides is often genuinely valuable—but how much preparation effort its inclusion requires, and whether that effort is justified given the specific application being built.

Market Data

FX rates, interest rates, volatilities, and commodity prices are some of the key required inputs for any AI application in exposure management, hedge recommendation, or portfolio optimization. This data is typically sourced from market data providers or directly from banking platforms. Market data is generally well-structured and available in real time. The treasury-specific considerations are around history depth, market conventions, rate type consistency (spot, forward, mid, bid/ask), and the handling of market disruptions that can distort model training if not addressed explicitly.

How AI Uses Treasury Data

Understanding where data comes from is one side of the readiness question. The other side is understanding what the AI actually does with it—because different AI applications use data in fundamentally different ways, and the data requirements are different for each.

Training Data: Teaching the Model

Training data is the historical dataset that an ML model learns its patterns from. The quality of training data is the most direct determinant of model quality: a model cannot learn patterns that are not present or are inconsistently represented in its training data, and it will faithfully learn patterns that are present even if those patterns are artifacts of data quality issues rather than genuine business signals.

For treasury forecasting, training data is the historical record of cash flows, positions, and the contextual features derived from them—day of week, proximity to payroll dates, month-end flags, and so on. For fraud detection, training data is the historical record of transactions, including whatever examples of genuine fraud exist in the dataset. For exposure consolidation, training data is the history of how exposures have been classified and how they have evolved over time.

Two practical considerations govern training data quality in treasury: the length of history available (generally, more is better, with a minimum of two years for seasonal patterns to be learnable) and the consistency of that history (breaks caused by system migrations, reorganizations, or policy changes need to be identified and handled rather than fed to the model as-is).

Inference Data: Operating in Production

Inference data is the live operational data fed to a deployed model to generate its predictions, recommendations, or classifications. It must match the format and characteristics of the training data closely enough that the model's learned patterns remain valid. A model trained on daily closing balances extracted from the TMS will not perform reliably if the production feed switches to intraday cumulative balances without adjustment.

This matching requirement creates a practical discipline: any change to a production data feed—a new field, a changed extraction logic, a different bank format, a system migration—must be assessed for its potential impact on inference quality before it is implemented. This is one of the reasons that data pipeline mapping (knowing exactly where every field comes from, how it is transformed, and what format it arrives in) is an operational necessity for production AI systems, not just a development-phase exercise.

Classification

Classification is the use of AI to categorize incoming data into discrete buckets. In treasury, the most common classification applications are transaction categorization (assigning a cash flow category to an incoming transaction), fraud screening (fraud versus not-fraud), and exposure classification (cash flow exposure versus balance sheet exposure, hedgeable versus non-hedgeable).

Classification models require training data in which the categories are clearly and consistently labeled. If the historical record contains categorization changes—the same transaction type classified differently at different periods, or categories that have changed definition over time—the classification model will learn those changes and reproduce them.

Anomaly and Outlier Detection

Anomaly detection is the use of AI to identify data points that deviate significantly from learned patterns. Unlike classification, which sorts data into predefined categories, anomaly detection works by learning what "normal" looks like and flagging what falls outside it. In treasury, this applies to fraud detection, outlier cash flows that may represent data errors or unusual business events, and model monitoring—detecting when incoming inference data is beginning to look different from what the model was trained on, which is an early indicator of model drift.

The data quality requirement for anomaly detection is somewhat different from that for classification or forecasting. Because the model is learning "normal," it needs training data that is genuinely representative of normal operations—without large volumes of anomalous transactions that would distort the baseline. Periods of unusual business activity (pandemic disruption, major restructuring, large one-off transactions) are worth isolating or excluding from anomaly detection training data for exactly this reason.

Data Mapping, Integration, and Guardrails for Production

Building a production AI system on top of treasury data requires more than a well-prepared training dataset. It requires a sustainable data pipeline—the infrastructure that moves data from its source systems to the AI on a reliable, monitored schedule. It should have several components:

Data pipeline mapping documents every field: where it comes from, what transformations are applied, what validation checks it passes through, and what the downstream AI application expects to receive.

This documentation is what makes the system auditable and maintainable. When a feed fails or a field changes, the map is what allows the problem to be diagnosed and resolved without reconstructing the logic from memory.

Integration testing before production deployment confirms that each data source is delivering what the model expects, in the format and frequency the pipeline assumes. Testing should cover not just the happy path—data arriving clean and on schedule—but also failure modes: what happens when a bank feed is late, when the TMS produces an unusual output, when a field is unexpectedly null.

Guardrails around data quality determine what the system does when incoming data fails a quality check. The options are a hard stop (the pipeline halts and alerts the team), a graceful degradation (the system operates on the most recent reliable data while flagging the issue), or a bypass (the affected data source is excluded while the rest of the pipeline continues). The appropriate choice depends on the criticality of the specific data source to the application's output. Defining these guardrails explicitly—before going live, not after the first production incident—is what separates a production-grade system from a prototype that was never designed to fail safely.

The Data Readiness Assessment

Before committing to any AI implementation, a structured data readiness assessment helps determine whether the data needed for a specific application is available in sufficient quantity and quality to support reliable model training. A practical assessment covers four dimensions:

Availability: Does the data exist in a structured, accessible format? Data that exists in reports but not in extractable tables, or that requires manual assembly from multiple sources, creates extraction challenges that are worth understanding before AI development begins.

History depth: Does sufficient historical data exist to train reliable models? Most forecasting applications benefit from a minimum of two years of daily data, with three or more years being preferable. For fraud detection, the requirement includes sufficient examples of both fraud and non-fraud cases—and in well-controlled organizations, genuine fraud examples are rare by design, which creates a class imbalance that requires specific handling.

Consistency: Has the data been recorded consistently over the intended training period? Changes in the chart of accounts, system migrations, policy changes, and organizational restructuring all create consistency breaks that are worth identifying and handling explicitly. A model trained across a consistency break may learn that break as a pattern.

Completeness: Are there missing values, gaps, or coverage holes that would affect model training? Daily cash position data with occasional missing days, bank statement feeds that experienced interruptions during a system migration, forecast data with gaps in subsidiary submissions—all of these benefit from explicit handling.

The output of this assessment is not a go/no-go decision. It is a clear picture of the data preparation work required while AI development begins—and a realistic estimate of how long that work will take. Both outcomes are useful: knowing what you are working with is always better than discovering it mid-project.

It is also worth noting that AI-assisted tools can accelerate several of the preparation steps that follow—data profiling, anomaly detection, and pattern-based cleansing in particular are areas where the same AI capability being applied to treasury decisions can be turned usefully toward the data preparation itself.

Data Preparation: What the Work Involves

The amount of data preparation required varies significantly from one organization to the next. A treasury team operating on a consistently maintained ERP system with clean historical data and well-documented bank feeds may find the preparation work straightforward. One working through a recent system migration or post-merger integration may have more to address. The data readiness assessment described in the previous section is what determines which situation you are actually in—and that is a more useful starting point than any general estimate of effort.

Extraction and integration—pulling data from multiple source systems into a unified dataset—is typically the first step. In a typical ERP or TMS environment, this means defining the right combination of tables, fields, and filters to capture relevant transactions, handling large data volumes efficiently, and building extraction logic that can be run repeatedly as new data is generated.

Normalization—bringing data from different sources into a consistent format and semantic structure—follows extraction. Bank transaction codes must be mapped to a consistent taxonomy. ERP or TMS document types must be classified into cash flow categories. Currency amounts must be converted to a consistent base currency at appropriate rates. Entity hierarchies must be applied consistently.

Cleaning—identifying and handling data quality issues—is where the most judgment is required. Some issues have clear resolutions: duplicate records can be deduplicated; obviously incorrect values can be corrected or excluded. Others require business judgment: how to handle months of inconsistent data following a system migration, whether to include or exclude a period of unusual activity, how to treat large one-off payments that may distort the distribution.

Feature engineering—creating the derived variables that ML models learn from—is the final data preparation step and the one most directly connected to domain expertise. Raw transaction data may not contain the signals that predict treasury outcomes directly. The signals emerge from derived features: day of week, days to month end, proximity to payroll dates, rolling averages of recent flows, ratios between forecast and actual for prior periods, binary flags for quarter-end and year-end.

This is where treasury domain expertise becomes a genuine advantage. Knowing that month-end AP concentration is a structural pattern, that payroll dates create predictable cash movements, and that quarter-end behavior differs meaningfully from mid-quarter—these are the kinds of insights that translate directly into better features and, in turn, better models. The combination of treasury knowledge translated into technical execution is what produces the best outcomes at this stage.

Building for Ongoing Data Quality

A data foundation requires ongoing maintenance as business conditions change, systems are upgraded, and new data sources are added. Building data quality management into the operational cadence from the beginning avoids the gradual divergence between training data and current reality that can quietly degrade model performance over time.

Automated data quality checks should run as part of every data ingestion cycle—verifying that expected data sources have delivered on schedule, that record counts are within expected ranges, that key fields have no unexpected null values, and that balance checks between systems are within acceptable tolerance. When checks fail, the failure should be visible promptly rather than discovered during the next model training run.

Data lineage documentation—recording where each data element comes from, how it was transformed, and what quality checks it passed—is important for both audit and troubleshooting. When an AI recommendation appears anomalous, the first question is usually whether the underlying data was correct. Data lineage documentation makes that question answerable efficiently.

Change management for data. Changes to source systems, organizational structures, or business processes that affect how data flows into an AI application are worth flagging, ideally before they are implemented, not after. The discipline of treating the AI system as a downstream stakeholder in change management—the same way treasury would flag a system change that affects a payment workflow—helps prevent silent data degradation over time.

A Practical Starting Point

The data foundation work described in this chapter is important—but it does not all need to happen before anything useful can be built. The right approach is to start with one well-defined application, work through the data preparation for that specific use case, and build on those lessons.

Cash flow forecasting for a single major operating account, or fraud detection for a single payment stream, are both good starting points. Work through the data readiness assessment, the extraction, the normalization, and the cleaning for that one application. The issues discovered and the solutions developed will prove relevant to every subsequent application.

The data work is, in one sense, never fully finished. Business conditions change, systems evolve, and the AI's data requirements grow as its scope expands. Organizations that treat data quality as an ongoing priority—investing in its consistency and accessibility as a matter of course—tend to find that each successive AI application is easier to implement and faster to produce value than the last. The data foundation, like the AI models that depend on it, compounds over time.

PUT IT INTO PRACTICE

Conduct a data inventory for your highest-priority AI use case.

Pick one treasury AI application you are seriously considering—cash flow forecasting, payment fraud detection, or FX exposure management. Then answer these questions for that specific application:

- What data sources would this application require? List every system—TMS, ERP, bank feeds, planning tools, legacy systems.
- For each source, does the data exist in a structured, extractable format?
- How far back does clean, consistent history extend?
- Are there known consistency breaks—system migrations, reorganizations, policy changes—in that history?
- Who owns each data source, and what would it take to get extraction access?
- Do you know what type of AI use this data would serve—training a model, feeding inference, supporting classification, or enabling anomaly detection? The answer shapes the specific quality requirements.

This inventory will give you a clear picture of what data preparation work is needed before AI development can begin. Either outcome is useful: a strong foundation confirms you are ready to move forward; identified gaps confirm where to focus first.

CHAPTER SUMMARY

- Treasury data is more complex than it may initially appear—it lives in multiple systems, reflects the history of organizational change over many years, and contains business context that the data itself may not capture.
- Treasury data sources span five key categories: the book of prime record and transactional data (ERP or TMS), banking data (BAI2, MT940, camt.053, SWIFT), planning and forecasting systems (dedicated platforms or Excel), legacy systems, and market data—each with specific characteristics that affect AI readiness.
- AI uses treasury data in four distinct ways: as training data to build models, as inference data to generate live predictions, as classification inputs to categorize transactions or exposures, and as anomaly detection baselines to surface deviations from learned norms—each with different data quality requirements.
- Production AI systems require data pipeline mapping, integration testing, and explicit guardrails around data quality failures—defining what the system does when data is late, incomplete, or outside expected ranges is a prerequisite for a production-grade implementation.
- A structured data readiness assessment across four dimensions—availability, history depth, consistency, and completeness—is a practical first step before any AI implementation begins.
- The data preparation effort varies by organization, depending on the state of existing systems and data quality; AI-assisted profiling and anomaly detection can accelerate the early stages,

while feature engineering—where treasury domain expertise is translated into technical execution—remains the step most directly connected to model quality.

- Ongoing data quality management, including automated checks, lineage documentation, and change management coordination, is most effective when built into the operational cadence from the beginning—and increasingly, AI tools can carry much of this monitoring work themselves.
- Starting with one application and one well-defined data source allows the lessons from data preparation work to compound into a stronger foundation for every subsequent implementation.

In the next chapter, as we begin PART III, "Governance and Control," we turn to one of the most important requirements for AI in treasury: explainability. When an AI agent makes a decision that affects your cash position, your hedge portfolio, or your payment operations, being able to explain that decision to your CFO, your auditor, and potentially your regulator. Building systems that are both intelligent and transparent is the challenge we address next.

PART III

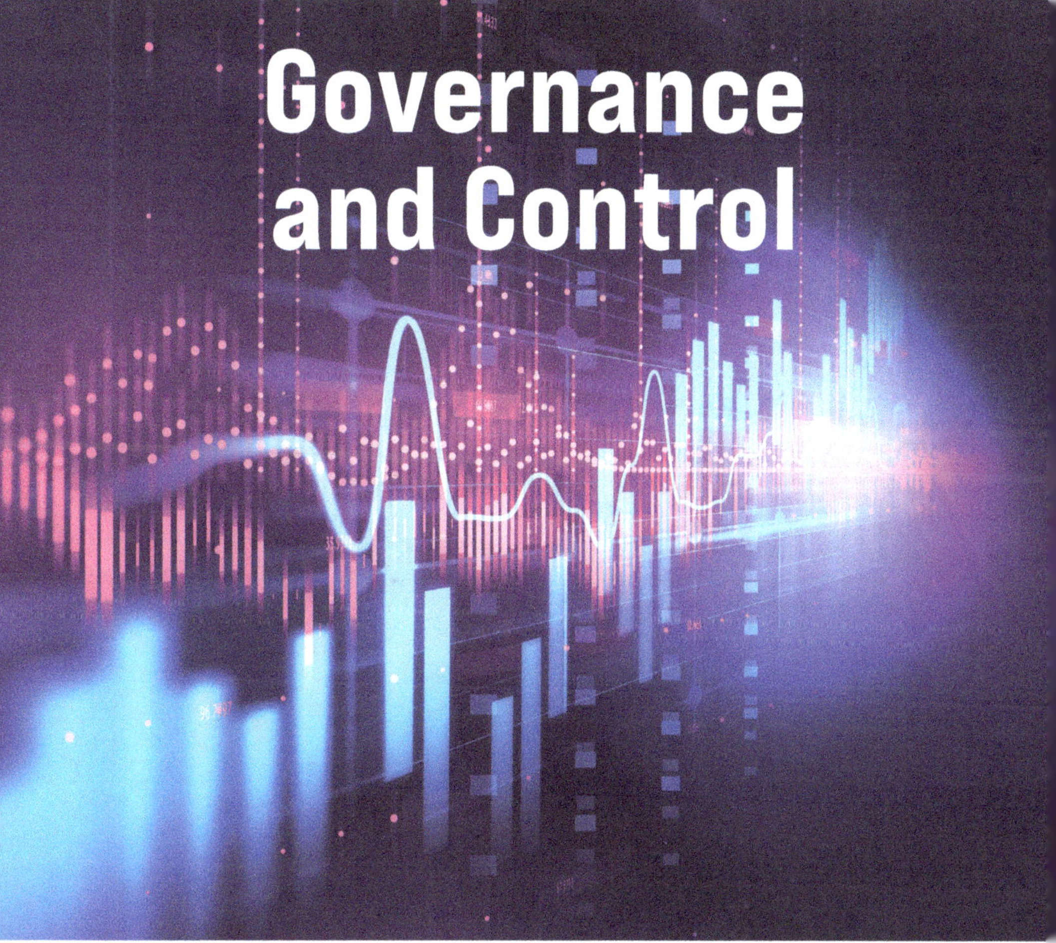

> **“Great responsibility follows inseparably from great power.”**
>
> **French National Convention**
> Committee of Public Safety (1793)

CHAPTER 9

Controls, Guardrails, and Explainability

Applying Proven Control Principles to Agentic Treasury Systems

The control principles that make financial systems trustworthy—segregation of duties, audit trails, validation checks, human override—apply directly to agentic AI systems. This chapter translates four proven control types into their AI equivalents, introduces the guardrail design that governs the FX exposure management pipeline as an example, and addresses the explainability that any AI system in treasury needs to satisfy for SOX compliance and CFO accountability.

In 2009, I co-wrote a book on SOX Compliance with SAP Treasury and Risk Management (SAP Press, First Edition). The central argument of that book was straightforward: sound internal controls in treasury are not a compliance burden imposed from outside—they are the architecture of a trustworthy system. The controls that satisfy an auditor are the same controls that protect the organization, enable confident decision-making, and create the audit trail that allows problems to be diagnosed and corrected when they occur.

More than a decade later, that argument applies directly to agentic AI in treasury—and in some ways with even greater force. When a human treasury professional makes a decision, the control framework around them is designed to prevent errors, detect anomalies, and create accountability. When an AI agent makes a decision, the same requirements apply. The technology is different. The control principles are not.

This chapter draws on those established principles—specifically the four types of controls that have governed treasury implementations for decades—and applies them to the design of guardrails for agentic AI systems. It also addresses explainability: the requirement, which is both practical and ethical, that AI-driven decisions in treasury must be explicable to the people who are accountable for them.

A note on the examples in this chapter: they draw heavily on SAP treasury and risk management implementations, and intentionally so. SAP's ERP architecture contains the most complete real-world realization of all four control types—systemic controls built into the platform, configurable controls set through the configuration menu, programmable controls developed through custom code and user exits, and manual controls enforced through approval workflows and reconciliation procedures. This full spectrum, embedded in a system that processes treasury transactions at scale across global organizations, makes it the most instructive reference for what these control principles look like in practice. The principles themselves, however, are universal. They apply with equal force to any treasury management system, any agentic AI architecture, and any financial system where accountability and auditability are requirements.

Why Controls Matter More, Not Less, With AI

There is a tempting assumption that AI systems, because they are automated and algorithmic, are inherently more controlled than manual processes. The assumption is understandable but wrong.

Manual processes are visible. When a treasury analyst makes a forecasting error or approves a payment incorrectly, the error is usually traceable: there is a spreadsheet, an email, a decision log. The control framework around manual processes is designed to catch these errors before they have consequences, and to identify them quickly when they do.

AI systems can fail in ways that are less visible. A model that has drifted and is producing systematically biased forecasts may not trigger any obvious error signal—the outputs look plausible, the system is running normally, and the degradation accumulates quietly until someone notices that decisions made on the AI's recommendations are consistently falling short. An agent that encounters an edge case outside its training data may produce a technically valid output that is operationally wrong—and the output will have no hesitation attached to it.

This is why the control framework for an agentic AI system must be at least as rigorous as the control framework for the manual processes it replaces, and in some respects more so. The stakes of an AI error are potentially larger, because AI systems operate at a speed and scale that human processes cannot match. When a payment fraud scheme evades an AI screening agent, each transaction isn't reviewed individually, often by different people; the AI is likely to process many transactions, using the exact same procedure, before the detection failure is identified.

The good news is that the control principles needed are well-established. They have been applied to SAP treasury systems for decades. The challenge is not inventing new controls—it is understanding how the same proven control types apply within an AI-based multi-agent architecture.

Before the mapping, a brief orientation. The four control types (preventive, detective, directive, and corrective) are well-established in financial systems governance, and each has a direct equivalent in an agentic AI system:

- **Preventive controls** stop an unwanted action before it occurs.
- **Detective controls** identify when something has gone wrong after the fact.
- **Directive controls** guide behavior through policy and instruction.
- **Corrective controls** address and resolve problems once identified.

The sections that follow explore how each of these control types maps onto its AI equivalent—systemic, configurable, programmable, and manual—and how each functions in the agentic context.

The Four Control Types, Applied to AI Agents

TABLE 9.1 maps each control type to its AI agent equivalent. The implementation mechanisms differ from their human-process equivalents—but the underlying purpose of each control type translates directly.

TABLE 9.1: Control types and their AI equivalents

Control Type	Nature	AI Agent Equivalent
Systemic/ Built-in	**Preventive**	Hard stops at agent hand-offs preventing invalid outputs from passing forward; balance integrity checks; automatic reconciliation between agent outputs
Systemic/ Built-in	**Detective**	Complete audit trail of every agent input and output; drill-down from decision to data; error logs; transaction logs at each pipeline stage
Configurable	**Preventive**	Orchestrator policy parameters: trade size limits, approved counter-party lists, hedge ratio ranges, payment approval thresholds—all configurable without code changes
Configurable	**Detective**	Configurable alert thresholds; variance limits that trigger human review; outlier sensitivity settings; model performance monitoring triggers
Programmable	**Preventive**	Custom validation rules between agents; workflow with approval triggers; interface with external validation systems (credit checking, sanctions screening)
Programmable	**Detective**	Exception reporting dashboards; model accuracy monitoring; bias direction tracking; agent performance benchmarks
Manual	**Preventive**	Treasury professional override mechanisms; human approval gates for decisions above threshold; policy documentation and acknowledgment
Manual	**Detective**	Daily review of override log; weekly variance analysis; period-c model accuracy review; external audit of AI decision trail

The four AI control types—systemic, configurable, programmable, and manual—should never operate in isolation. A robust control framework combines all four, creating multiple and compensatory layers. The same principle that governs ERP control design governs agentic AI control design: no single control is sufficient, and the failure of any one layer should be caught by another.

Systemic Controls: The Non-Negotiables

Systemic controls are built into the system by design. In an SAP ERP environment, they include things like document posting integrity, automatic synchronization between main and sub-ledgers, and the fundamental rule that posted documents can be reversed but not deleted. These are not configurable options—they are architectural commitments that define the trustworthiness of the system.

In an agentic AI system, the equivalent systemic controls are the non-negotiable design principles of the multi-agent architecture described in Chapter 4. They should be treated as foundational, not as features that can be switched off for convenience.

Agent Output Integrity

Each sub-agent in the pipeline must produce a structured, validated output before passing work to the next sub-agent. A sub-agent that produces a malformed, incomplete, or internally inconsistent output should not be permitted to pass that output forward. The orchestrator's validation checkpoints at each handoff are the AI equivalent of SAP's document posting integrity check—they ensure that what enters each stage of the pipeline is what the next stage expects.

In practice, this means defining the schema for each sub-agent's output—what fields it must contain, what ranges are valid, what checks must pass before the output is accepted—and building those validation checks into the orchestrator rather than relying on each receiving sub-agent to detect problems with its inputs.

Complete and Immutable Audit Trail

In SAP treasury, the principle that posted documents can be reversed but not deleted is a systemic control that creates an immutable record. Every change has a timestamp, an author, and a document trail. This principle must carry over directly to agentic AI systems.

Every agent action—every data extraction, every calculation, every recommendation generated, every compliance check performed—should be logged with its inputs, outputs, timestamp, and the version of the model or rule set that produced it. This log should be append-only: entries can be added, but not modified or deleted.

Retrofitting an audit trail onto a system that was not designed to produce one is significantly harder than building it in from the start—and an audit trail added after the fact is inherently less credible than one that has been running since day one. The audit trail is not a reporting feature to be added when the system is otherwise complete. It is a foundational design requirement.

When an auditor asks, "Why did the system recommend this hedge trade?", the answer should be traceable from the final recommendation back through the compliance check, the trade sizing calculation, the netting result, and the data extraction that started the pipeline. Not "the AI decided" but a complete, documented chain of reasoning.

Automatic Reconciliation Between Agents

SAP's automatic synchronization between main and sub-ledgers ensures that accounting entries in the general ledger match the details in subsidiary ledgers without requiring manual reconciliation. The equivalent in an agentic AI system is automatic reconciliation between the outputs of adjacent agents.

If the data extraction agent reports 47 open FX exposures and the netting calculation agent produces results for only 44, something has been dropped. That discrepancy should be caught automatically by the orchestrator, not discovered by a treasury professional reviewing the final recommendations and noticing that some exposures appear to be missing.

Configurable Controls: Policy as Parameter

Configurable controls are those that an organization implements by choosing from a set of available options. In SAP treasury, these include transaction execution rules, limit management, payment blocks, dual controls for sensitive access fields, and release controls. They require no custom programming—they are configuration decisions that translate the organization's treasury policy into system behavior.

In an agentic AI system, configurable controls live in the orchestrator. This is one of the most important architectural principles of the multi-agent design: policy should be centralized in the orchestrator as configurable parameters, not distributed across individual agents as hardcoded logic.

Limit Management and Approval Thresholds

The orchestrator should enforce the same limit structure that the treasury policy specifies for human decision-making. A payment that requires CFO approval above $5 million in the manual process should require CFO approval above $5 million when generated by an AI agent. A hedge trade that exceeds counterparty credit limits should be blocked, whether a human dealer or an AI agent generates it.

The critical design principle is that these limits are parameters in the orchestrator—not logic embedded in individual agents. When the policy changes—the threshold moves from $5 million to $3 million, a new counterparty is approved, a currency is added to the restricted list—the update happens in one place. Every agent in every pipeline that touches that control automatically reflects the change.

Segregation of Duties Between Sub-Agents

Segregation of duties is one of the most fundamental controls in treasury operations. The person who initiates a payment should not be the person who approves it. The person who manages a bank relationship should not have sole authority to execute transactions with that bank. These separations exist because no individual should have unchecked control over a complete transaction cycle.

In an agentic AI system, the equivalent principle is segregation of responsibilities between sub-agents. The sub-agent that generates a hedge recommendation should not be the sub-agent that executes the trade. The sub-agent that calculates the net FX exposure should not be the sub-agent that checks whether the resulting hedge satisfies compliance requirements. Each sub-agent should do exactly one thing, and the sub-agent that validates should be architecturally independent from the sub-agent that generates.

This is not just good architecture—it is the control principle of segregation of duties expressed in AI design. An orchestrator that routes outputs through independent validation agents before allowing execution is implementing the same protection that dual authorization workflows provide in manual treasury processes.

Payment Blocks and Hard Stops

In SAP treasury, payment blocks prevent disbursements from processing until specific conditions are met. They can be set manually or triggered automatically by system conditions. The AI equivalent is a hard stop in the orchestrator: a condition that, when met, halts the pipeline entirely and requires explicit human action to restart.

Hard stops should be reserved for conditions that represent genuine risk of material error or policy violation—not used so liberally that they become routine interruptions that practitioners learn to dismiss. Calibrating the threshold for a hard stop versus a warning versus an informational alert is a judgment call that requires treasury expertise to set appropriately. The technology can enforce whatever threshold is specified; specifying the right threshold is a human responsibility.

TABLE 9.2 illustrates what these checks could look like in practice, drawn from the orchestrator validation layer between the netting calculation sub-agent (Agent 2) and the trade processing sub-agent (Agent 3) in the FX exposure pipeline described in Chapter 7. In the table, HALT conditions stop the pipeline entirely; WARN conditions require human confirmation; Filter conditions apply policy rules without interrupting the pipeline.

TABLE 9.2: Orchestrator validation checks between Agent 2 (Netting Engine) and Agent 3 (Trade Processing) in the FX exposure pipeline

Check	What It Verifies	If It Fails
Spot Rate Freshness	Rates must be less than 4 hours old	HALT—reload rates from TCURR/Bloomberg
Data Quality	Netting run must return CLEAN status	HALT—return to Agent 2 for reprocessing
Checksum Verified	Secure Hash Algorithm (SHA)-256 checksum must match Agent 2 registry	HALT—potential data tampering, security alert
Required Currencies	EUR, GBP, JPY must appear in every run	HALT—missing currencies may indicate extraction error
Entity Count	All 6 company codes must be in netting pool	WARN—treasurer must confirm before proceeding
Materiality Screen	Only expose positions above USD 500K to Agent 3	Filter—sub-threshold positions excluded
Variance Check	Exposure must be within ±20% of prior month	WARN—flag for treasurer review

The three response types reflect the calibration principle: HALT for conditions that represent genuine data integrity or security risk, WARN for conditions that warrant human judgment before proceeding, and Filter for policy rules that apply automatically without stopping the pipeline. The specific thresholds—four hours for rate freshness, 20% for variance, USD 500K for materiality—are configurable parameters in the orchestrator, not hardcoded logic in any individual agent.

Programmable Controls: Custom Intelligence

Programmable controls require custom development. In an SAP ERP context, these include ABAP function modules, user exits, custom validation rules, workflow with email triggers, and audit table updates. They are the controls that address requirements specific to the organization's business that standard system configuration cannot cover.

In an agentic AI system, programmable controls are the custom validation logic, anomaly detection rules, and cross-agent consistency checks that go beyond what standard orchestrator configuration provides. They are where the organization's specific risk profile and business complexity are encoded into the control framework.

Custom Validation Rules

Some validation requirements are specific enough to the organization's business that they cannot be expressed as generic threshold parameters. A treasury operation that manages exposures across multiple regulatory jurisdictions may need validation logic that checks whether a proposed hedge satisfies the specific documentation requirements for each jurisdiction's accounting treatment. A payment operation with a complex counterparty network may need validation logic that checks for patterns suggesting collusion or override of fraud controls.

These custom validations sit between agents in the orchestrator's validation layer—not inside any individual agent. This preserves the single-responsibility principle of the agent architecture while allowing the control framework to be as sophisticated as the business requires.

Anomaly Detection as a Detective Control

The fraud detection and outlier identification capabilities described in Chapter 3 are examples of programmable detective controls. They do not prevent an anomalous transaction from being processed; they identify it after it has been processed (or before it is executed, in a real-time screening context) and surface it for review.

In an agentic AI system, anomaly detection should run continuously, not just at specific pipeline stages. The pattern of agent outputs over time—the distribution of hedge recommendations, the frequency of compliance flags, the variance between forecast and actual—should itself be monitored for anomalies. A sudden change in the pattern of outputs may indicate a data quality problem, a model drift event, or an unexpected change in business conditions. Each of these requires different responses; all of them require early detection.

Audit Table Updates and Logging

In SAP ERP, programmable controls include audit table updates that record changes to sensitive data. The equivalent in an agentic AI system is the comprehensive event log that records not just agent outputs but the context in which they were produced: the input data, the model version, the configuration parameters in effect, and the validation checks that ran.

This log is what makes the system auditable under SOX. The question "did the system operate within its defined controls when it generated this recommendation?" should be answerable from the log without requiring a retrospective investigation. The log is not an afterthought—it is a primary output of the system.

Manual Controls: The Human Layer

Manual controls are developed and implemented internally by the organization. In SAP treasury, they include approvals and authorizations, policies and procedures, daily and monthly reconciliation, and the review of clearing accounts. They are not inferior to automated controls—they are the human judgment layer that automated controls cannot fully replace.

This is perhaps the most important design principle for agentic AI in treasury: manual controls are not a fallback for when automation fails. They are a permanent, essential layer of the control framework. The question is not whether humans should remain in the control loop for AI-driven treasury processes, but which decisions require human involvement and at what frequency.

The Override as a Control Mechanism

The ability for a treasury professional to override an AI-generated recommendation is not a concession to human skepticism about AI. It is a control mechanism with the same standing as any other manual control. The treasurer who overrides a hedge recommendation because they know the underlying exposure is about to change is exercising exactly the judgment that the manual control layer is designed to capture.

Overrides should be logged with the same rigor as agent outputs: who applied the override, when, based on what decision, with what reasoning. Over time, the pattern of overrides is itself a control metric: frequent overrides in a specific category may indicate that the model's assumptions in that area need updating, that the data inputs are unreliable, or that the policy parameters do not reflect current business conditions.

Periodic Review as a Detective Control

Manual detective controls—the periodic review that looks across a period of AI-generated decisions and asks whether they are collectively consistent with policy and good judgment—are a necessary complement to real-time automated monitoring. Automated controls catch individual anomalies. Periodic human review catches patterns that emerge only over time: a systematic bias in hedge sizing, a consistent timing difference between AI recommendation and human approval, a gradual drift in the distribution of decisions that no individual flag triggered.

The frequency and scope of periodic review should be defined in the governance framework—which is the subject of Chapter 10—and should evolve as the organization builds confidence in the system. Early in a deployment, weekly reviews of AI-generated recommendations alongside the actuals they predicted may be appropriate. As confidence builds, monthly reviews may suffice for stable processes, with more frequent reviews reserved for high-risk areas.

Periodic review should also include active testing of the guardrail system itself—not just reviewing what the agents have been doing, but deliberately testing whether the guardrails hold. The analogy from SAP treasury implementations is relevant: a control that has never been tested is a control you cannot rely on. The three types of testing translate well into the AI context: negative testing, boundary testing, and scenario testing.

Negative testing asks each agent to do something it should refuse: submit a trade recommendation that exceeds the single-trade size limit, pass a payment instruction for a counterparty not on the approved list, attempt a netting run with a currency missing from the required set. Each of these should be blocked or escalated by the guardrail system.

If any agent complies with a request it should have refused, the relevant policy rule needs to be tightened before the next production run.

Boundary testing probes the edges of permitted behavior: a trade sized at exactly the limit, then one unit above it; an exposure variance at exactly the permitted threshold, then just beyond it. The system should behave consistently and predictably at the boundary—not produce different results depending on how closely a limit is approached. Inconsistent boundary behavior is often a sign that a policy rule has been implemented in more than one place, and the implementations differ slightly.

Scenario testing runs the full pipeline through realistic positive and negative cases: a clean netting run that should produce trade recommendations and proceed smoothly; a run with a deliberately stale rate feed that should halt at the spot rate freshness check; a run where one legal entity is missing from the netting pool, which should trigger a warning requiring human confirmation. Documenting the expected outcome for each scenario before running it—and comparing the actual outcome against the expectation—creates the testing record that demonstrates that the control framework is operating as designed. This is the AI equivalent of the access reviews and control testing that treasury teams already conduct under established governance frameworks. And like those access reviews, the results should be documented, signed off on, and retained as evidence of ongoing control effectiveness.

Explainability: The Treasury Requirement

The concept of explainability in AI refers to the ability to describe, in terms a human can understand, why an AI system produced a particular output. In many AI applications, explainability is desirable but not essential. In treasury, it is a practical requirement with direct regulatory and governance implications.

The Auditor's Question

SOX compliance requires that financial reporting controls be documented and that their effectiveness be tested. When an AI system makes a decision that affects financial reporting—a hedge recommendation that determines accounting treatment, a payment approval that releases funds, a cash positioning decision that affects liquidity disclosures—the auditor's question is not "did the AI decide correctly?" It is "did the system operate within its defined controls, and can you demonstrate that?"

This question requires more than accuracy. It requires documentation of the decision logic, evidence that the control framework was operating as designed, and a clear line from the decision back to the data and rules that produced it. A system that produces correct outputs but cannot explain how it produced them does not satisfy a SOX audit, regardless of its accuracy.

Explainability at the Agent Level

The multi-agent architecture described in Chapter 4 supports explainability by design. Because each agent does exactly one thing and produces a structured output, the explanation of any final recommendation can be constructed by tracing the pipeline: the data the extraction agent retrieved, the net positions the netting engine calculated, the hedge recommendation the trade processor generated, the compliance checks the approval agent ran, and the result of each.

This is structurally different from the explainability challenge of a monolithic AI model, where the same question—why did the model recommend this trade?—requires opening the black box and inspecting internal model weights. The multi-agent architecture makes the explanation a natural by-product of the audit trail, not a separate technical exercise.

Plain Language Explanation

For explainability to serve its governance purpose, the explanation must be understandable to the people who are accountable for the decision. When a CFO asks why treasury recommended a specific hedge, the answer needs to be in treasury language: the exposure that triggered the recommendation, the policy ratio that determined the sizing, the counterparty that was selected and why, the accounting treatment that was applied.

The Key Drivers approach from the forecasting POC in Chapter 6 illustrates this principle: factor importance is expressed in business language rather than statistical terms, with each driver traced back to a business reality that the treasurer can verify or challenge. This same principle should govern the explanation design for every agent pipeline in a production treasury AI system.

A Practical Note on Implementation

The control framework described in this chapter is comprehensive by design—it is meant to be complete, not minimal. In practice, not every element needs to be implemented before a system can go into production. The right approach is to identify the controls that are most critical for the specific use case being deployed, implement those rigorously, and extend the framework incrementally as the system's scope expands.

For a first treasury AI deployment—say, a cash forecasting system where the AI generates recommendations and humans make all final decisions—the systemic controls (audit trail, output integrity) and manual controls (override logging, periodic review) are the minimum viable control framework. Configurable controls become more important as the system moves toward autonomous action. Programmable controls become more important as the use case becomes more complex.

The 90-day pilot framework in Chapter 11 is designed to be run within this control framework from day onc. Starting with appropriate controls, even if minimal, builds the right habits and creates the audit trail that will be needed when the system's scope and autonomy expand.

PUT IT INTO PRACTICE

Assess your current treasury AI control framework—or, if you are planning an implementation, the control framework worth building.

- Map your existing treasury controls to the four types: systemic, configurable, programmable, and manual. For each, ask whether the equivalent control exists in your AI systems or implementation plans.
- Identify your highest-risk control gap. Where in the AI pipeline is the control framework weakest relative to the risk of the decision being made? Start there.
- Define your audit trail requirements before implementation begins. What questions must the system be able to answer for an auditor? Work backwards from those questions to define what the log must capture.
- Review your segregation of duties design in the agent architecture. Is the agent that generates recommendations architecturally independent from the agent that validates them? If not, this is a design issue worth addressing before deployment.
- Define your override policy. Who can override AI recommendations, under what conditions, and with what documentation requirement? The override policy should be written and acknowledged before the system goes live.

CHAPTER SUMMARY

- Sound controls for agentic AI in treasury are not a new invention—they are the application of proven control principles from financial systems to an AI-based multi-agent architecture; the technology changes, the principles do not.
- AI systems can fail in ways that are less visible than manual process failures; the control framework must be at least as rigorous as for the processes AI replaces, and in some respects more so.
- The four control types—systemic, configurable, programmable, and manual—all apply to agentic AI systems, with specific equivalents for each: output integrity checks, orchestrator policy parameters, custom validation rules, and human override mechanisms.
- Segregation of duties between sub-agents—the agent that generates a recommendation should be architecturally independent from the agent that validates it—is the AI equivalent of one of treasury's most fundamental control principles.
- Policy should live in the orchestrator as configurable parameters, not distributed across individual agents as hardcoded logic—this is what makes policy changes surgical rather than scattered.
- The immutable audit trail—every agent action logged with its inputs, outputs, timestamp, and model version—is the foundation of SOX compliance for AI-driven treasury decisions.
- Manual controls are a permanent layer of the framework, not a fallback—the override mechanism, periodic review, and human approval gates are controls with full standing alongside automated controls.
- Explainability in treasury is a governance requirement, not a technical nicety—the multi-agent architecture supports it by design, because the audit trail of agent handoffs is the explanation.

Chapter 10 turns to the broader governance, risk, and compliance framework within which these controls operate—the organizational structures, accountability mechanisms, and regulatory considerations that make AI-enabled treasury operations trustworthy at the institutional level, not just the technical one.

CHAPTER 10

Governance, Risk, and Compliance

Building the Institutional Framework for AI-Enabled Treasury

Technical controls are necessary but not sufficient. This chapter addresses the organizational framework that makes AI in treasury institutionally trustworthy: accountability structures, the three lines of defense, SOX compliance for automated controls, and the autonomy graduation process—how trust in AI systems is built incrementally and deliberately. It also covers how to respond when the unexpected occurs.

The previous chapter addressed the technical control framework for agentic AI systems: the guardrails, audit trails, and validation mechanisms that make individual AI decisions trustworthy. This chapter addresses something broader and equally important: the organizational and institutional framework within which those technical controls operate.

A technically sound control framework is necessary but not sufficient for trustworthy AI in treasury. The questions of who is accountable for AI decisions, how AI policy is set and maintained, how the organization demonstrates compliance to external parties, and how trust in AI systems is built incrementally over time—these are governance questions that cannot be answered by technical design alone. They require deliberate organizational choices.

This chapter draws on two bodies of established practice. The first is the three lines of defense model, which has governed risk management and internal control in financial services for decades. The second is the SOX compliance framework for treasury systems, which provides specific guidance on what documentation, testing, and accountability structures are required when treasury processes are automated. Both frameworks predate AI by many years. Both apply directly to AI-enabled treasury operations, with specific adaptations for the nature of AI decision-making.

Accountability: Who Owns the AI Decision?

The first governance question in any AI implementation is accountability. When an AI agent makes a recommendation—a hedge trade, a payment approval, a cash positioning decision—who is responsible if that recommendation is wrong?

The answer, in my view, is clear: the treasury professional who reviewed and approved the recommendation is accountable for the decision. This is not a concession to AI skepticism. It is a principle that follows directly from how accountability works in any system involving automated processes and human oversight.

When a payment runs through SAP's automatic payment program and is released, the treasury professional who set up the payment run and authorized its execution is accountable for that payment. The fact that the system processed it automatically does not transfer accountability to the software. The same principle applies when an AI agent generates a recommendation that a treasury professional approves: the professional who approved it is accountable.

This accountability principle has an important corollary: it only holds best when the treasury professional brings genuine judgment to the review. The governance design should make that judgment easy to apply—providing the right information, at the right level of detail, in the time available—rather than creating review processes that are so fast or so cursory that they add friction without adding insight. The goal is a review process that is genuinely useful to the person conducting it, not one that exists primarily to satisfy a control requirement.

For decisions made autonomously by AI agents—within pre-approved parameters, without individual human review of each decision—the accountability question is answered differently. The treasury leader who defined the parameters and approved the scope of autonomous operation is accountable for the class of decisions within those parameters. This is another reason why the policy-setting process for AI autonomy deserves careful attention.

The Three Lines of Defense, Applied to AI

The three lines of defense model provides a well-established framework for organizing risk management and internal control responsibilities. The three lines are operations, risk and compliance, and internal auditing. It has been applied to treasury operations for decades and translates directly to AI-enabled treasury, with specific adaptations for each line, as shown in **TABLE 10.1**.

First Line: Operations

In traditional treasury, the first line of defense is the treasury team itself: the professionals who execute transactions, monitor positions,

TABLE 10.1: The three lines of defense, applied to AI

Line of Defense	Traditional Treasury	AI-Enabled Treasury
First—Operations	Treasury team: execution, monitoring, exception management	AI agents for routine execution; treasury professionals for oversight, overrides, and exception judgment
Second—Risk and Compliance	Risk management and compliance functions: policy setting, limit monitoring, regulatory reporting	AI policy framework and guardrail design; compliance validation agents; regulatory reporting automation; AI performance monitoring
Third—Internal Audit	Internal audit: independent testing of control effectiveness	Independent testing of AI control framework; audit of AI decision trails; assessment of model performance and drift; validation of override patterns

and manage exceptions. The first line is closest to the risk and has the primary responsibility for operating within policy.

In an AI-enabled treasury, the first line is a combination of AI agents and treasury professionals. The agents handle routine execution within defined parameters; the professionals provide oversight, manage exceptions, and exercise judgment in situations the agents are not designed to handle. The control responsibilities of the first line do not diminish with AI adoption—they shift. The treasury professional's first-line responsibility moves from executing routine tasks to overseeing the agents that execute them.

This shift has a practical implication for team design: the treasury professionals in an AI-enabled operation need to understand how the AI systems work at a level sufficient to identify when they are behaving abnormally. The first-line oversight responsibility requires the capability to exercise it.

Second Line: Risk and Compliance

The second line of defense includes the policy framework, limit structures, and compliance monitoring that the first line operates within. In traditional treasury, those responsible for this function set hedging policies, monitor limit utilization, and ensure that treasury operations comply with applicable regulations.

In an AI-enabled treasury, the second line has an expanded role: it is also responsible for the AI policy framework itself. This includes defining what decisions AI agents are permitted to make autonomously, what parameters govern their behavior, and what monitoring is required to ensure ongoing compliance. The second line's responsibility extends from monitoring treasury operations to monitoring the AI systems that conduct them.

This is a new capability requirement for second-line functions. A risk and compliance professional who can evaluate traditional treasury controls but has no framework for evaluating AI control design is not equipped to fulfill the second-line responsibility in an AI-enabled operation. Building this capability—either through hiring, training, or external support—should be part of any AI implementation governance plan.

Third Line: Internal Audit

Internal auditing provides independent assurance that the control framework is effective. In a SOX context, this includes testing controls, documenting results, and providing the evidence that management's assessment of control effectiveness is well-founded.

For AI-enabled treasury operations, the internal audit's scope must extend to the AI control framework. This means testing whether the guardrails designed in Chapter 9 are operating as designed, whether the audit trail is complete and immutable, whether override patterns are consistent with policy, and whether the AI systems are performing within their documented parameters.

Auditing AI systems requires some adaptation of traditional audit techniques. Testing that a payment approval workflow operates correctly is well-established in practice. Testing that an AI recommendation engine is producing recommendations consistent with its defined policy parameters requires a different kind of analysis. Internal audit functions that will be responsible for AI systems need to develop these capabilities, and the AI teams that build the systems need to design them with auditability as a first-class requirement.

The audit community itself is adapting, and treasury professionals can help shape that process. Those who have worked through major system implementations will recognize the pattern: auditors trained on legacy architectures sometimes ask for documentation or reconciliation reports that an integrated system simply does not produce in the form they expect—because the separation of records that required reconciliation no longer exists in the same way. AI systems will create similar learning curves. The audit trail an AI system produces is different in structure from a manual process trail—more comprehensive in some respects, unfamiliar in others. Building a productive working relationship with internal and external auditors before the first AI audit, rather than during it, is time well spent. Walking auditors through how the system works, what the audit trail contains, and what questions it can and cannot answer in advance of a formal review turns a potentially difficult conversation into a constructive one.

SOX Compliance Implications

For organizations subject to SOX—US-listed companies and many others who voluntarily apply equivalent standards—the automation of treasury processes through AI raises specific compliance considerations that are worth addressing directly.

What SOX Requires for Automated Controls

SOX requires management to assess the effectiveness of internal controls over financial reporting (ICFR) and to obtain independent auditor attestation of that assessment. When automated controls—including AI-driven processes—affect financial reporting, those controls fall within scope.

The key SOX requirement for any automated control is that it must be documented, tested, and demonstrably effective. For a traditional SAP automated control—an automated posting rule, a system-enforced approval threshold—this is well-established practice. The documentation describes what the control does; the test confirms it is operating as described; the evidence is available for the auditor.

For an AI agent that influences financial reporting—through hedge recommendations that determine accounting treatment, through payment approvals that affect cash and payables balances, through forecasts that inform liquidity disclosures—the same requirements apply. The control must be documented (what parameters govern the agent, what validation checks run, what human review occurs), tested (does the system behave as documented in both normal and exception conditions), and demonstrated to be operating effectively.

The Documentation Requirement

In my experience with SOX compliance in SAP treasury environments, documentation quality is often the difference between a smooth audit and a difficult one. The controls may be operating correctly, but if they are not documented in a way that allows an auditor to understand what they do and verify that they are doing it, the audit becomes more challenging than it needs to be.

For AI controls, the documentation requirement is more demanding than for traditional automated controls, because the behavior of AI systems is less deterministic and more dependent on data and model state. The documentation for an AI control should include the following: what the agent does and what parameters govern its behavior; what data it relies on and how data quality is ensured; what validation checks the orchestrator runs on its outputs; what human review occurs and how it is documented; and how the system's performance is monitored over time.

This documentation works best as living documentation—updated when parameters change, when models are retrained, and when the scope of autonomous operation expands.

In a well-designed multi-agent system, much of this documentation work can be handled by a dedicated documentation sub-agent. Rather than relying on the team to update documentation manually after each change, the documentation agent monitors the system automatically—capturing parameter changes, logging model retraining events, recording override decisions and their outcomes, and maintaining a timestamped audit trail of how the system has evolved. It can also generate structured summaries of agent behavior on demand, which is useful both for internal governance reviews and for preparing audit evidence. The documentation sub-agent does not eliminate human judgment about what matters and what to record, but it removes the friction that causes documentation to fall behind in practice.

Change Management and SOX

SOX controls are not just evaluated at a point in time. They are evaluated over the course of the year, and any significant changes to controls during the year require disclosure and may require additional testing. For AI systems, this creates a specific consideration: model retraining and parameter updates are operational events that may constitute changes to automated controls.

A practical approach is to define, in advance, which changes to the AI system constitute material changes to the control framework—requiring formal documentation, change testing, and potentially disclosure—and which are routine operational updates that do not. A parameter change that moves a trade size threshold from $5 million to $3 million is likely a material change to the control. A model retrain that uses the same architecture and parameters on updated data is likely an operational update. The line between them is worth defining in the governance framework before changes begin occurring.

Building the AI Policy Framework

The AI policy framework is the document—or set of documents—that defines what AI systems in treasury are permitted to do, under what conditions, with what oversight, and with what accountability. It is the foundation on which the control framework of Chapter 9 and the governance structures of this chapter rest.

What an AI Policy Framework Should Cover

A practical AI policy framework for treasury should address the following at minimum:

Scope of autonomous operation. Which treasury decisions may be made autonomously by AI agents, within what parameter ranges, and for what transaction types? Which decisions always require human review regardless of parameter compliance? The scope should be specific—not "routine treasury decisions" but "payment approvals below $X for approved counterparties within approved payment methods."

Escalation criteria. What conditions trigger escalation to human review? These should be defined as specific thresholds and conditions, not general principles. An AI agent that encounters a compliance flag should know exactly who to escalate to and what information to provide. Escalation paths that depend on human judgment to determine should be designed out of the system.

Override authority. Who may override AI recommendations, at what level of decision, and with what documentation requirement? Override authority should mirror the authorization structure for equivalent manual decisions—if a manual payment above $5 million requires dual authorization, an override of an AI recommendation that affects a payment above $5 million should require the same.

Performance standards. What accuracy and reliability standards must AI systems meet to remain in production? What is the threshold for taking a system offline for retraining or review? Who makes that determination, and what process governs it?

Review cadence. How frequently are AI system performance, override patterns, and compliance with policy parameters reviewed? By whom, with what documentation, and with what authority to make adjustments?

Who Sets AI Policy

AI policy for treasury is a treasury leadership responsibility. The parameters that govern what AI agents can do are, in effect, the organization's treasury policy expressed in executable form. They carry the same authority as any other treasury policy and should be approved at the same level.

In practice, this means that the CFO or treasurer who approves the treasury policy manual also approves the AI policy framework, at least for the parameters that translate treasury policy into AI agent behavior. The technical team that builds the AI systems can advise on what is technically feasible; they cannot substitute for the treasury judgment that determines what is appropriate from a business standpoint.

This accountability boundary—technical teams determine what is possible based on business needs, treasury leadership determines what is permitted—should be explicit in the governance structure. It is a form of segregation of duties applied at the organizational level.

Regulatory Considerations

The regulatory environment for AI in financial services is developing. As of this writing (mid 2026), explicit regulatory guidance on AI use in treasury operations is limited, but several existing regulatory frameworks have direct implications that are worth noting.

What Is Already Clear

The principle that automated controls over financial reporting must be documented, tested, and demonstrably effective applies regardless of whether those controls involve traditional software or AI—this is established SOX practice that extends naturally to AI systems.

The principle that financial institutions must be able to explain significant decisions to regulators applies regardless of whether those decisions were made by humans or AI systems. This is not an AI-specific regulation; it follows from existing supervisory expectations around model risk management and internal controls.

Regulations governing specific treasury activities—hedge accounting under IFRS 9 and ASC 815, payment operations under various national frameworks, foreign exchange controls in specific jurisdictions—apply to AI-assisted execution of those activities just as they apply to manual execution. An AI-generated hedge designation that does not meet the documentation requirements of IFRS 9 is non-compliant regardless of its technical accuracy.

What Is Still Developing

Specific regulatory guidance on AI governance in treasury is still emerging. Several regulatory bodies have published principles-based guidance on AI risk management in financial services—the direction is clearly toward greater scrutiny of AI decision-making, model validation requirements, and human oversight standards. The specific requirements for treasury applications have not yet been fully defined.

The practical implication is that organizations that build explainability, documentation, and governance into their AI systems from the start will be better positioned to demonstrate compliance with whatever specific requirements emerge. Retrofitting governance onto a system that was designed without it is significantly harder than building it in from the beginning.

Building Trust Incrementally

Trust in AI systems is not granted by policy—it is earned by performance. This is the same principle that governs any change in how treasury operations are conducted: new processes, new systems, and new approaches earn the confidence of the organization through demonstrated reliability, not through assertion.

The governance framework should reflect this reality by building in explicit mechanisms for trust to develop and be validated over time, rather than assuming that a technically sound system will automatically earn operational trust.

The Autonomy Graduation Process

When an AI system is first deployed, the appropriate starting point is typically the most conservative end of the Autonomy Spectrum—recommendations only, with human approval of every decision. As the system demonstrates reliable performance over a defined period, the governance framework should provide a clear process for expanding the scope of autonomous operation.

This graduation process should be explicit: what performance standard must the system demonstrate, over what time period, verified by whom, before the scope of autonomous operation expands? The process should not depend on informal agreement or the enthusiasm of advocates—it should be documented, governed, and verifiable.

The autonomy graduation process is also the mechanism through which the organization builds the institutional knowledge needed to govern AI systems effectively. The treasury professionals who oversee an AI system in its recommendation-only phase are developing the understanding of how it behaves that will allow them to exercise meaningful oversight when its autonomy expands.

When to Pull Back

The governance framework must include explicit criteria for reducing or suspending AI autonomy, not just for expanding it. A system that encounters unexpected conditions, produces anomalous outputs, or shows signs of model drift should have a clear mechanism for human review and intervention that does not require a crisis to trigger.

The most common failure mode in AI governance is that the criteria for pulling back are less well-defined than the criteria for expanding autonomy. Organizations that thought carefully about the conditions under which AI would earn more independence often have not thought as carefully about the conditions under which it would earn less. Both deserve equal attention in the governance framework.

PUT IT INTO PRACTICE

Assess your organization's AI governance readiness.

- Does your treasury have a documented AI policy framework, or are AI tools currently operating on informal understandings about what they are permitted to do? If the latter, defining the policy is the first governance priority.
- Who in your organization has accountability for AI decisions in treasury? Is this clearly defined and understood by the people involved? If the accountability structure is ambiguous, clarify it before expanding AI autonomy.
- Has your internal audit function assessed the AI control framework? If AI systems are affecting processes within SOX scope and have not been audited, this is a gap worth closing.
- Does your second-line risk and compliance function have the capability to evaluate AI controls? If not, what do they need to develop that capability—training, external support, or additional expertise?
- Do you have a documented process for expanding and contracting AI autonomy over time? If the governance framework only addresses the initial deployment and not how the system's scope evolves, it needs revisiting.

CHAPTER SUMMARY

- Accountability for AI decisions in treasury rests with the treasury professional who act on the recommendations—the same principle that governs accountability for automated processes in any financial system.
- The three lines of defense model—operations, risk and compliance, and internal audit—applies directly to AI-enabled treasury, with specific adaptations for each line's expanded role.
- SOX compliance requires that automated controls affecting financial reporting be documented, tested, and demonstrably effective—this applies to AI agents just as it applies to traditional automated controls.
- Change management for AI systems—distinguishing material control changes from routine operational updates—must be defined in advance to manage SOX implications of model retraining and parameter updates.
- The AI policy framework should cover the scope of autonomous operation, escalation criteria, override authority, performance standards, and review cadence—approved by treasury leadership, not delegated to technical teams.
- Regulatory guidance on AI in treasury is still developing; organizations that build explainability and governance in from the start are better positioned to demonstrate compliance as requirements emerge.
- Trust in AI systems is earned through demonstrated performance, not granted by policy—the governance framework should include explicit, documented processes for both expanding and contracting AI autonomy over time.

PART IV turns from governance to implementation—Chapter 11 sets out a 90-day pilot framework that offers a structured approach to getting AI into production in a treasury environment, with appropriate controls from day one.

PART IV

Implementation

> **“To know what you know, and to know what you do not know — that is knowledge.”**
>
> **Confucius**
> *The Analects*, 2.17

CHAPTER 11

Making It Real

A Practical Guide to Your First AI Implementation

This chapter is a practical guide to getting from interest to implementation. It draws on the lessons I learned while building both proofs-of-concept—the cash forecasting system and the FX hedge management system—to describe what a well-sequenced first AI initiative involves: choosing the right problem, doing the foundational work before building, and validating in the right order. The people dimension—how roles evolve and how AI literacy develops—is addressed as part of the implementation.

Treasury professionals are no strangers to major implementations. System migrations, ERP upgrades, TMS deployments, in-house bank implementations, payment factory buildouts—these are complex, multi-phase projects that treasury teams have navigated successfully for decades. They involve data migrations, change management, parallel runs, phased go-lives, and the particular challenge of maintaining operational continuity while transforming the underlying systems.

AI implementation in treasury sits within that broader context. In some organizations, it will follow a major system transformation—adding an intelligence layer on top of a newly cleaned and consolidated data architecture. In others, it will run alongside ongoing system work, addressing specific pain points that do not need to wait for a full transformation. And in some cases, the AI initiative will itself catalyze the data quality and integration work that the organization has been deferring.

What distinguishes AI implementation from traditional treasury technology projects is not the complexity of the project management—treasury teams are well-equipped for that—but the nature of the outputs and the speed at which value can be delivered. Traditional treasury technology projects are typically long-cycle transformations: ERP implementations, TMS deployments, and payment factory buildouts measured in months or years. AI initiatives can be structured very differently. A well-scoped AI proof of concept can produce working results on real data in weeks. That difference in pace changes how you plan, how you measure success, and how you build organizational confidence. An AI system does not produce a report or execute a workflow in the way a TMS or ERP system does. It learns from data, generates estimates and recommendations, and improves over time. Understanding how to scope, test, and validate that kind of output is genuinely new for most treasury organizations, and it is what this chapter addresses.

The framework here uses a 90-day first initiative as its working example—an illustration, not a prescription. The right timeline depends on the use case, the state of the underlying data, the organization's existing infrastructure, and the complexity of the integration required. Some initiatives will be faster. Others, particularly those requiring data remediation first, may take longer. The principles are the same regardless of the timeline.

Choosing the Right First Initiative

The most consequential decision in any AI implementation is which problem to solve first. A well-chosen first initiative creates momentum, builds organizational confidence, surfaces challenges early, and establishes a foundation for what follows. A well-chosen one creates momentum and builds the foundation for what follows; a less well-chosen one makes the next attempt harder to build on.

When I set out to build the cash forecasting proof of concept described in Chapters 5 and 6, and later the FX exposure management system described in Chapter 7, these were the questions I worked through. They apply to any treasury AI initiative.

Start with a real pain point, not an interesting problem. One related principle that the FX build reinforced: choose a problem where you already know what a good solution looks like. The goal of any AI-enabled initiative is to accelerate and enhance a process you already understand—not to use AI to solve something that has resisted solution by other means. If you cannot describe what success looks like before the build starts, measuring it afterward becomes very difficult.

The pain point I was addressing in the forecasting POC was one I had observed consistently in treasury operations at large multinational organizations: cash flow forecasting was being done with significant manual effort and limited accuracy. Treasury teams were assembling daily positions from multiple sources, maintaining forecast models in spreadsheets anchored to last year's actuals, and operating with limited visibility beyond a week. The problem was structural and widely shared—and notably, it persisted even in organizations running tightly integrated ERP systems such as SAP, where the data exists but forecasting has historically required significant manual assembly, customization, and interpretation on top of it. That made it a strong candidate—not because it was the most technically interesting challenge, but because solving it would be immediately visible and valuable to the people doing the work.

In the next section, "Before You Begin: Three Foundations," I'll provide step-by-step instructions, but I want to foreground that discussion with some suggestions to keep in mind as you figure out where to start, gleaned from my experience with developing the two proofs of concept:

Map the data before anything else. Before any architecture was designed or any code was written for the forecasting POC, I mapped the data. The core sources were transaction history from the TMS and ERP system—AR wire receipts, AR ACH credits, the AR lockbox, AP disbursements, payroll, debt service, tax payments, and intercompany flows—plus bank statement feeds for daily closing positions. The critical design decision was to forecast at the category level rather than at the aggregate closing balance, because category-level forecasting reveals the composition of any variance and points directly to the appropriate response. An aggregate shortfall tells you that something requires attention. A category-level breakdown tells you what to do about it.

Choose the architecture for the problem, not for the technology. The decision to use a three-model ensemble—ARIMA for short horizons, XGBoost for medium, Prophet for longer—came from observing that different forecast horizons have fundamentally different statistical characteristics. No single model serves all three well. The architecture followed from the problem. The same principle applied to the FX system: five agents, each doing one thing cleanly, were employed, rather than a single system attempting all five, because the FX pipeline has five distinct responsibilities that map to five distinct points of control—mirroring the segregation of duties principle that has governed sound financial systems for decades.

Define the success criterion before building anything. The specific question the forecasting POC was built to answer was testable from the start: does a three-model ensemble, weighted by horizon, produce meaningfully better forecasts than a single Prophet-only baseline? A testable question means a clear definition of success from day one. Initiatives that begin without this tend to discover their success criteria late, making it harder to demonstrate value and know what to build upon.

Design the output for the user, not for the technology. The forecasting dashboard was built as a Streamlit application with seven tabs, each serving a distinct purpose in a treasury workflow. Every feature was written in treasury language. The confidence level indicator reflects the model's internal uncertainty in terms that a cash manager can act on. The Key Drivers tab explains the forecast in terms of business factors, not model weights. If the person using the system cannot understand why it is telling them what it is telling them, the system will not be used.

For most treasury organizations, AI-assisted cash flow forecasting for a single major operating account is a strong first initiative for the same reasons the POC focused there: the data is typically available, the success criterion is measurable, the output is reviewed daily, and the scope is completable in a focused first phase.

Before You Begin: Three Foundations

Before the implementation work starts, three foundations need to be in place: a written definition, a focused data assessment, and a governance structure. None takes long to establish, but all of them affect the likelihood of a successful outcome.

A Written Definition

Write the following down, in one page or less:

- What process the initiative will address
- What data it requires and whether that data exists in an accessible form
- What the measurable success criterion is
- Who the initiative owner is
- Who the executive sponsor is
- Who the technical lead is

Once the document is complete, get the owner, sponsor, and technical lead to agree to the written version.

The discipline of writing this down creates the shared understanding that prevents scope creep, expectation mismatch, and the accountability gaps that derail technology implementations. It is the single page that makes everything else move faster.

A Focused Data Assessment

In the cash forecasting POC, the data assessment came before any modeling work. I pulled two years of daily transaction history, categorized by flow type, and tested it for consistency. What I found was instructive: there were three months where AP categorization had shifted following an ERP configuration change, and a period of intercompany activity reflecting a one-time restructuring. Both needed explicit handling before training—not because the data was bad, but because those periods would have been learned as patterns if fed to the model uncritically.

The data assessment does not need to be exhaustive before you start. It needs to be honest. Pull a sample, look at what is there, identify the breaks and gaps, and estimate the preparation effort. That estimate will shape your timeline more than any other single factor. One practical mechanism worth noting from the cash forecasting POC: the Settings tab in the dashboard was designed to allow the treasury user to register known future events—scheduled debt service, tax payments, and other one-off items—that the model cannot predict from history alone. This structured input mechanism bridges data-driven forecasting with the domain knowledge the treasury professional holds, and it is part of the data design work, not an afterthought.

A Governance Structure

Three roles are sufficient for a focused first initiative: an initiative owner (the treasury professional accountable for the success criterion and primary user of the outputs), a technical lead (who owns the data pipeline and model development, and needs enough treasury understanding to ask the right questions), and an executive sponsor (who provides organizational support at key milestones without needing day-to-day involvement). Weekly thirty-minute checkpoints keep a focused initiative on track.

What Building an AI Treasury System Actually Looks Like

The architecture principles described in Chapters 4 and 7—sub-agents, single responsibility, orchestrator guardrails—may sound abstract until you see what they look like in a working build. This section describes the implementation of the FX exposure management system from the practitioner's perspective: the decisions made, the tools used, and what the process of building actually feels like. None of this requires a technical background. It requires knowing what you are commissioning.

The Build Is Not Linear

The FX exposure management system is not being built from Agent 1 through to Agent 5 in sequence. The approach is frontend-first. The first thing to establish is the dashboard—what a treasury user would actually see. Before a single line of agent logic was written, we established what the output needed to look like: the netting summary, the trade recommendations, the compliance flags, the audit trail. Starting from the user's view forced every subsequent technical decision to be made in the right direction.

The netting engine follows, refined from bilateral to multilateral logic, with the IHB account ledger and spot rate conversion mechanics added iteratively. The orchestrator guardrail system follows that, built one check at a time, each tested against realistic mock data before the next was added. The agents follow the guardrails, not the other way around.

This sequencing—user output first, then logic, then controls—is the opposite of how traditional treasury technology projects are typically structured, where the system is built, and then the user interface is added at the end. For AI implementations, the reversal is deliberate. You understand what you build. You trust what you can validate against your own domain knowledge. Starting from the output you need keeps every build decision grounded in the business problem.

Phase Zero: The Work That Makes Everything Else Possible

Most AI implementations skip a critical step. They move straight to the visible work—building the netting engine, designing the trade processor, wiring the dashboard—because that is where progress is tangible, and momentum is easy to sustain. Even though you start by establishing what will be on the dashboard, the foundational work that should come first produces nothing you can demonstrate on a screen, and so it tends to get deferred or skipped entirely.

The consequence shows up later. Six months into the build, someone asks where the hedge ratio policy is documented, and the answer is that it exists in three places, and is in a slightly different form in each. A decision is made to add a sixth agent, and this requires refactoring the orchestrator, which means touching everything. The implementation has been building on sand without knowing it.

Before any agent is built, three things need to exist: a single, locked data source that every agent draws from; the policy configuration file written as the system's constitution—every rule, every parameter, every escalation threshold in one place; and a clear definition of what each agent will receive and what it will pass on. These are not technical artifacts. They are the functional expert's decisions, written down. The technical team cannot make them. Only the treasury professional who understands the policy, the data, and the required output can.

This phase is unglamorous by design. It produces a folder structure, a written policy document, and a definition of data flows—nothing that runs, nothing that renders on a screen. But the architecture now has what might be called load-bearing discipline: policy lives in one place, every agent knows what it receives and passes on, and a change to the hedge ratio or the approved counterparty list propagates automatically rather than requiring a manual search through the codebase. Every decision from this point becomes easier because these constraints exist.

The same principle applies to when an external technical partner is engaged—specifically, the point at which the AI system needs to connect to a production environment. Moving from a proof of concept on representative data to a live system requires integrating with a production ERP system, TMS, or banking platform: real endpoints, real security models, real data flows. That integration work belongs to specialists who understand the platform's connectivity layer. The most effective handover happens after the mock-data pipeline is working end to end and has been validated against treasury policy—not before. At that point, the brief to the technical team is precise: here is what the first agent expects to receive from the production system, and here is what the final agent produces for it to consume; please wire both ends securely into the live environment. Engaging them earlier, before the functional logic is stable, means building on shifting ground. Engaging them later means hand-coding integration work they should own. The timing of that engagement is one of the most important project management decisions in an AI implementation—and it is a functional judgment, not a technical one.

The Orchestrator as Policy Document

One of the clearest insights we got from building the FX system is that the orchestrator is not primarily a technical component. In Claude Code (my LLM of choice for the POCs), we defined the orchestrator in a file called a policy configuration file (a plain text file with an .md extension—

readable by both humans and the system, requiring no specialist software to edit). Every agent operates under these instructions. Every guardrail check runs against these rules. If a policy parameter changes—the hedge ratio moves from 75% to 80%, a new counterparty is approved, a currency is added to the restricted list—you update one line in the policy configuration file. All agents reflect the change automatically.

This is the key insight for treasury professionals who are thinking about AI implementation: you do not need to be a developer to design an AI-enabled treasury system. The functional expert needs to be able to articulate the policy rules precisely. What instruments are approved? What are the counterparty limits? What hedge ratio is required? What conditions trigger escalation to the group treasurer rather than proceeding automatically? If you can answer these questions in writing, you can define an orchestrator. The technical team builds the infrastructure. The functional expert writes the policy. That is the right division of labor.

There is one risk worth naming explicitly at this stage: the functional knowledge that defines how the system operates—why the hedge ratio is set at a particular level, why specific currencies require escalation, why certain counterparties are on the approved list—can easily end up living only inside the system's operating instructions, with no separate record of the reasoning behind each decision. For a production system in a risk-sensitive treasury environment, that creates a genuine audit and continuity risk. The policy configuration file is the executable version of the policy; a short functional specification document—written in plain language, reviewed by the treasury team, and maintained alongside the system—is what makes that policy defensible to an auditor and recoverable if the person who built it moves on.

One thing the FX build is making clear: the functional rules and policy decisions you define for your agents—the thresholds, the escalation criteria, the approved parameters—are as important as any governance document, and they carry the same obligation to be kept current. A well-designed multi-agent system can include a dedicated documentation agent that handles much of this automatically, generating audit trails and maintaining a running record of how the system operates. The goal is a system that can explain itself.

What an Agent Pipeline Might Look Like

TABLE 11.1 shows a basic five-agent FX management pipeline using SAP as an example—each agent with a single responsibility, a defined input, and a defined output. Nothing passes between agents without the orchestrator validating it first.

What **TABLE 11.1** illustrates is not the FX workflow specifically—that is covered in Chapter 7. What it illustrates is the architecture discipline. Each agent receives a structured input, does one thing with it, and passes a structured output. The orchestrator sits above the pipeline, validating each handoff. If the data extraction agent returns an unexpected result, the pipeline halts at that point. The problem is isolated, diagnosable, and correctable without touching any other agent.

TABLE 11.1: An example five-agent FX management pipeline

Agent	Single Responsibility	Receives	Passes On
1.Data Extraction	Reads IC transactions from the ERP or TMS via the platform's provided API	Company codes, date range	26 raw IC line items
2. Netting Engine	Applies multi-lateral netting—one net per currency per entity	26 raw IC line items	Net position per entity per currency + IHB FX exposures
3. Trade Processing	Sizes FX hedge, selects instrument and tenor, applies IFRS 9 rules	IHB net FX exposures + spot rates	Trade recommendations (FX forward per currency)
4. Approval and Compliance	Checks policy limits, SOX controls, counterparty credit limits	Trade recommendations	Approved or escalated trades
5. SAP Execution	Creates approved FX deals in SAP TRM via BTP OData	Approved trades	Deal confirmations posted to SAP TRM

It is worth noting that a production FX hedge management implementation will typically involve significantly more agents than this example shows. The full lifecycle—from exposure extraction through netting, reconciliation, hedge recommendation, policy validation, approval routing, trade execution, SAP capture, settlement, accounting, and compliance monitoring—maps naturally to twenty or more specialized agents, each applying the same single-responsibility principle. The five-agent example illustrates the architecture discipline; a production build extends it.

The Technical Stack in Plain Terms

A treasury professional evaluating or commissioning an AI implementation does not need to know how to build with these tools. They do need to know what they are and what role each plays. The following describes the stack we are currently using for the FX POC—many other tools and frameworks are available, and the right choice depends on your infrastructure and technical team. What matters is understanding what each layer does, so you know what you are commissioning:

Claude Code—the AI agent framework where the Orchestrator and agents are built and run. A command-line tool that can read codebases, write and execute code, and call external systems autonomously.

Python—the programming language in which the agents are written. Each agent is a Python module with a defined input schema and output schema.

SAP BTP OData—the standard SAP API layer for S/4HANA—used here because the POC was built on SAP. Other ERP and TMS platforms deliver equivalent connectivity through their own standard APIs: Oracle via REST/SOAP services, Kyriba and ION via their published integration layers, and so on. In every case the principle is the same: connect through the standard, approved, supported API layer, not around it. IT and Information Security teams can approve this architecture because it uses the access patterns the platform already sanctions.

FastAPI—the backend framework that exposes the agent pipeline as API endpoints, so the frontend dashboard can trigger runs and display results.

Streamlit (prototype) / React (production)—the dashboard layers. Streamlit is Python-native and fast to build, used for prototyping and validation. React is the intended enterprise production interface for a production deployment. The FX POC is being prototyped in Streamlit and validated against treasury policy and realistic mock data. Both POCs remain controlled proof-of-concept explorations—not live production systems.

One additional capability worth understanding is MCP—Model Context Protocol, Anthropic's standard for connecting Claude to external tools. Think of it as a universal connector: any tool that implements the MCP standard can be connected to Claude immediately. For the FX system, the current architecture uses direct Python-to-BTP-to-SAP integration (for SAP environments). MCP servers become relevant in the next phase, when the AI system can autonomously reach into the ERP or TMS during a conversation—for SAP, asking it to pull today's IC positions and run the netting analysis on demand, without a separate pipeline trigger. The BTP layer already in place is the right foundation for SAP environments; a future MCP server sits on top of it rather than replacing it. Other ERP and TMS platforms have equivalent API foundations that the same pattern applies to.

It is worth noting that the MCP pattern is already being adopted beyond treasury. Organizations building AI-powered workflows in other domains are using MCP servers to give their agents direct, structured access to internal data sets—removing the need for custom integration work each time a new data source is added. The experience from those implementations reinforces the same principle the POCs demonstrated: the agent architecture is sound, and the connectivity layer is what determines whether a proof of concept becomes a production system. As MCP adoption matures, the path from proof of concept to production for treasury AI systems is likely to become considerably shorter.

Questions Worth Asking When Evaluating AI Tools

At some point during a first initiative or broader AI program, a treasury team will need to evaluate AI tools, approaches, or implementation options—whether built internally or sourced externally. These questions cover the areas that most often reveal whether a tool is genuinely fit for purpose.

About the technology:
What type of AI does this use — ML, NLP, LLM, or a combination? What data does the model learn from — our own data, data from other organizations, or publicly available data? How often does the model retrain, and how is model drift handled over time?

About accuracy:
What is the false positive and false negative rate for fraud detection? What forecast accuracy is achieved in practice, at what horizon, and against what baseline? Are there performance metrics from live deployments that can be shared?

About limitations:
What types of cases does this handle poorly? How does the system fail, and what happens operationally when it gets something wrong? What level of human oversight is recommended and why?

About explainability and audit:
Can the system explain why it made a particular decision, in treasury language? What audit trail exists for AI-generated recommendations and actions? How have organizations using this system addressed SOX audit requirements?

About implementation:
What data is required, in what format, and over what historical period? What does a realistic implementation timeline look like, and what typically takes longer than expected? What is the ongoing maintenance requirement once the system is live?

Clear, specific, evidence-based answers indicate a well-understood system. Vague or deflective answers suggest either an immature tool or an incomplete picture of production performance.

Validation Is the Work

Building the system is the smaller part of the effort. Validating it is the larger part, and it cannot be skipped or deferred. In any AI treasury implementation, every agent should be run against realistic data before the next one is built. Every guardrail check should be triggered deliberately—not just the happy path, but the failure modes: what happens when a data feed is stale, when a processing run returns a quality flag, when a proposed action exceeds a defined limit.

In the FX POC, this is the discipline being applied: each check is tested before the system moves to the next stage. A guardrail system is only trustworthy if you have tested what it actually does when it fires.

Knowing how the system fails is as important as knowing how it succeeds. A system that has only been tested on clean data in normal conditions is not production-ready. The edge cases—the missing bank feed, the ERP configuration change that shifts a categorization, the one-off intercompany transaction that distorts the pattern—are exactly what will happen in production, and they are what the guardrails and override mechanisms need to handle.

A Framework for the First Phase

A focused first initiative typically moves through three stages: constructing the foundation, building and validating, and production and learning. The stages are sequential because each creates the foundation for the next.

Stage 1: Foundation

The objective is a working data pipeline and a baseline model producing outputs—not yet in production, but running on real data. The work is primarily data work: extraction, normalization, feature engineering, and handling of consistency breaks. This stage almost always takes longer than the initial estimate—building in extra time for it at the outset pays dividends later. The baseline accuracy metric—what the current process achieves without AI—is worth documenting at this stage. That clarity makes everything that follows easier to validate.

Stage 2: Build and Validate

The objective is a system of production quality, with real users reviewing real outputs. Decide on the components of the user interface before constructing the agents. Define the orchestrator policy in plain language before any agent logic is written. Then build and validate each agent against realistic data, testing failure modes explicitly alongside normal operation. The initiative owner begins working with outputs alongside the existing process; both systems are maintained. Their live feedback is what separates a model that validates well in development from one that earns operational trust.

Stage 3: Production and Learning

The objective is a system in production that users rely on, with a documented understanding of what it does well, what it does less well, and what comes next. The override mechanism is built, tested, and explained to every user before go-live. The final weeks are as important as the first: measuring accuracy, documenting what was learned, and identifying the three highest-priority improvements for the next phase will produce the organizational confidence that will make the subsequent initiatives faster and more impactful than the first, pilot project.

The People Dimension

A working AI system is not the end of the implementation—it is the beginning of a different kind of work. The organizations that extract lasting value from AI in treasury are those that think deliberately about how roles evolve, how capabilities are built, and how the team's relationship with the technology develops over time.

How Roles Are Evolving

The most common concern when AI is introduced into a treasury team is whether roles will disappear. The honest answer, based on every technology transformation the profession has been through, is that roles shift more reliably than they disappear. ERP systems did not eliminate treasury professionals; they changed what treasury professionals do. The same pattern is observable with AI. The shift is consistent: away from assembly and toward judgment, from execution to oversight, from formatting outputs to interpreting them. (See **TABLE 11.2**.)

A new profile is emerging at the intersection of financial expertise and AI system management. Identifying and developing these individuals early is one of the highest-return talent investments an organization can make.

Building AI Literacy

AI literacy in treasury does not mean writing code. It means understanding AI well enough to be an effective user: knowing when to trust the output, when to question it, and when to override it. Four practical capabilities matter most:

Understanding what the system is doing. Not the mathematical details, but the logic. The forecasting system discussed in Chapters 5 and 6 combines three models weighted by horizon. This anomaly detection system learned what normal looks like from 18 months of payment history. This conceptual understanding is what allows users to recognize when something looks wrong.

TABLE 11.2: The shift to agentic AI

Role	Today	With Agentic AI
Cash Manager/ Analyst	Building and maintaining spreadsheet mod-els; manually assembling the daily position from multiple sources	Reviewing AI-generated positions; applying overrides with documented reasoning; interpreting anomalies; focusing on exception management and decision support
FX/Risk Analyst	Manually pulling expo-sure reports; calculating hedge requirements; formatting and keying trade orders	Reviewing AI-generated hedge recommendations; exercising judgment on market timing and counterparty selection; managing relationship and documentation quality
Treasury Operations	Processing manual exception queues; reconciling positions across systems; chasing confirmations	Overseeing automated reconciliation; managing exception escalations; monitoring system health and data quality
Head of Treasury	Reviewing and approving routine decisions across the full process	Setting AI policy and guardrails; reviewing performance metrics; focusing on strategy and governance rather than execution approval
Treasury Analyst (emerging)	Administrative and reporting tasks	AI system configuration; data quality management; model performance monitoring; bridging treasury and technical teams

Knowing when to override—and why. The override is the most important human–AI interaction in any treasury system. Used well, it contributes information no model has access to—the customer who called to say their payment will be late, the internal policy change that hasn't yet propagated. An override used well—based on knowledge the model does not have—makes the system more accurate over time and contributes directly to the audit trail.

Reading accuracy metrics. Accuracy metrics, directional bias, variance trends, and model drift signals are operational feedback that treasury professionals need to interpret to manage AI systems responsibly. An afternoon spent understanding what these metrics mean for your specific system will do more for system reliability than most technical improvements.

Asking good questions. Treasury teams that can ask substantive questions about accuracy, failure modes, explainability, and data requirements can evaluate whether AI tools are genuinely fit for purpose. Those capabilities make it possible to evaluate tools on their actual demonstrated merits.

Identifying Your AI Champions

In almost every treasury team, there are one or two individuals naturally drawn to understanding how systems work—who go beyond the user interface, who experiment with new tools, who translate technical concepts for colleagues. Identifying them early and giving them time to engage deeply with AI implementations is the highest-return talent investment in any AI transformation.

The Leadership Responsibility

The AI policy questions described in Chapter 10—what decisions AI systems can make autonomously, under what conditions, with what oversight—are leadership decisions requiring treasury judgment, not technical expertise. The treasurer who delegates these entirely to a technology team is abdicating a governance responsibility. Treasury leaders who engage visibly with AI systems, review forecast accuracy, examine override logs, and hold regular discussions about where AI is delivering value set a tone that cascades through the team.

What Success Looks Like

A successful first initiative produces five things:

- A working system in production—generating real outputs that real users rely on daily
- A measured improvement against the documented baseline—demonstrable, auditable, and attributable to the AI system
- A clear and honest understanding of limitations—what works, what works less well, where human judgment remains essential
- An engaged and informed user group—treasury professionals who understand how the system works and have a basis for trusting its outputs
- A roadmap for the next phase—the three highest-priority improvements, the next use case, and any organizational changes needed for a broader program

A first initiative that produces all five is a success regardless of whether the accuracy improvement was modest or significant. The organizational learning and the foundation for what follows are often more valuable than the immediate performance gain.

PUT IT INTO PRACTICE

Write down the definition of your first initiative before anything else—a single page covering the following:

- What specific treasury process will this initiative address?
- What data does it require, and is there confidence that data exists in an accessible, consistent form?
- What is the measurable success criterion?
- Who is the initiative owner?
- Who is the executive sponsor?
- Who is the technical lead?

If all six questions can be answered clearly in writing, the foundation is in place. If any is unclear, that is where to focus first. The discipline of shared written agreement—before implementation begins—is the single most reliable predictor of a good start.

CHAPTER SUMMARY

- Treasury professionals bring deep project experience to AI implementation—the new element is understanding how to scope, test, and validate outputs that learn from data rather than executing fixed rules.
- Choose the right first initiative: start with a real pain point, map the data first, match the architecture to the problem, define a testable success criterion, and design outputs for the actual user.
- The cash forecasting and FX exposure POCs illustrate these principles: both were grounded in operational pain points observed in global corporate treasury, with architecture decisions following from the problem rather than the technology.

- Three foundations should be established before starting: a written one-page definition, a focused data assessment that surfaces consistency breaks early, and a governance structure with three roles.
- Build frontend-first: start with what the user will see, then build the agents, then add the guardrails. The orchestrator is a plain-language policy document—functional expertise translated into executable rules, not code.
- Validation is the work: every failure mode must be tested deliberately before production. A system tested only on clean data in normal conditions is not production-ready.
- Roles in treasury are shifting from assembly to judgment, from execution to oversight; the emerging profile at the intersection of treasury expertise and AI system management is the highest-return talent investment.
- A successful first initiative produces five things: a working system, a measured improvement, a clear understanding of limitations, an engaged user group, and a roadmap—the organizational learning is often as valuable as the performance gain.

The next chapter looks ahead—at where this technology is heading, what the trajectory means for treasury professionals, and what the practitioners reading this book should consider going forward.

CHAPTER 12

Looking Ahead

This chapter offers a practitioner's honest assessment of where treasury AI may be heading—not predictions, but observations about what is maturing, what is still developing, and what the direction of travel suggests for the profession. It closes with five observations for the practitioners doing this work: start somewhere specific, invest in data, be honest about what you are still learning, bring your team along, and share what you discover.

Making specific predictions about AI is a reliable way to look foolish in retrospect. The technology is moving faster than any particular forecast can track, and the gap between what is demonstrated in a research setting and what is reliably production-ready in a complex enterprise environment has consistently been wider than early expectations suggested.

That said, some trajectories in treasury AI are becoming clear enough to discuss honestly. Not as predictions, but as observations about where the early evidence is pointing—what is maturing, what is still emerging, and what remains genuinely distant. This chapter offers that assessment, in the same spirit as everything else in this book: a practitioner's honest read.

What Is Becoming More Reliable

Several applications of AI in treasury are moving from proof of concept to production-capable—not uniformly across all organizations, but in enough real environments that the question is shifting from "can this work?" to "what does it take to make it work here?"

Cash flow forecasting. The evidence from controlled explorations, including the POCs described in this book, suggests that ensemble approaches can deliver meaningful accuracy improvements at most forecast horizons for organizations with reasonably clean historical data. The open question is not whether the approach works in principle—the architecture is sound—but whether organizations can build and maintain the data pipelines that production systems require. Data quality and consistency, not modeling technique, will determine outcomes for most treasury organizations. The impact also compounds when forecasting is built collaboratively with other departments rather than owned entirely by treasury. Organizations that have done this well find that treasury's standing across the business changes—from a function that asks for information to one that helps the business make better decisions with it.

Payment screening and fraud detection. AI-based anomaly detection in payment flows is among the most mature applications in treasury. The pattern-recognition problem is well-suited to machine learning; the training data (historical payments) is typically available in structured form; and the failure mode (a missed fraud) is clearly defined and measurable. Straight-through processing rates and false positive rates are improving in production deployments. The governance question—who reviews flagged transactions, with what authority, under what time pressure—is where most of the remaining implementation complexity lives.

Reconciliation and position management. The daily position assembly and reconciliation tasks that occupy a significant share of treasury operations time are well-suited to AI automation. The inputs are structured, the rules are definable, and the output—a confirmed position—is verifiable. The multi-agent architecture described in this book is already being applied to these workflows in early production deployments. The constraint, again, is data quality: an automated reconciliation built on inconsistent feeds produces answers that look reliable, but are less trustworthy than those derived from a slower manual process.

Intercompany netting. The FX exposure management system described in Chapters 7 and 11 are representative of where intercompany automation is heading: rule-intensive, high-volume, low-tolerance-for-error workflows where agent-based automation can apply systematic logic at a scale human processes cannot match. The prerequisite work that needs to happen first—organizations where netting rules exist primarily as institutional knowledge rather than documented policy face a higher implementation cost.

What Is Still Emerging

A second tier of applications is technically feasible and demonstrably promising in controlled settings, but not yet reliably production-grade across the range of organizational complexity found in large multinationals.

Autonomous FX execution within defined parameters. The pipeline from exposure identification through netting calculation to trade recommendation is achievable now. The step from recommendation to autonomous execution—where the AI system places trades without individual human approval of each—requires a level of regulatory clarity, governance maturity, and demonstrated system reliability that most organizations have not yet reached. The technical architecture exists. The organizational and regulatory foundation is still being built.

Cross-entity treasury orchestration. Managing cash, FX, and payments across a large multi-entity structure—with the AI coordinating flows between entities, optimizing intercompany lending, and managing the consolidated group position in real time—is directionally where the technology is heading. The complexity of the problem is substantial: entity-specific regulatory constraints, tax implications of intercompany flows, multiple banking relationships and payment formats, and the coordination requirement across treasury, tax, and finance. Early deployments in simpler structures are producing encouraging results. Full-complexity multinational implementations remain a multi-year horizon.

Hedge accounting documentation automation. The documentation requirements under IFRS 9 and ASC 815—formal designation, effectiveness assessment, prospective and retrospective testing—are candidates for AI-assisted automation. The rules are defined, the data is available in the TMS, and the output is structured. The constraint is the combination of high compliance stakes and the specificity of regulatory interpretation. This application will mature as AI-generated output demonstrates consistent compliance across auditor review—something that is beginning to be seen in simpler hedge designations.

What Remains Genuinely Distant

Honest assessment requires naming what is not close, regardless of what vendor demonstrations or technology press coverage might suggest.

Full autonomous treasury. A treasury function where AI systems manage cash, FX, payments, and risk without systematic human oversight in the decision loop is not a near-term reality for complex organizations. The governance frameworks, regulatory requirements, and institutional trust needed to support this level of autonomy are not yet in place—and building them appropriately will take years of incremental, demonstrated performance.

AI that understands organizational context without being told. The AI systems available today are powerful at pattern recognition and rule application. They do not inherently understand why a particular subsidiary consistently forecasts conservatively, what the CFO's unspoken priorities are heading into a board meeting, or when a banking relationship requires a phone call rather than an automated instruction. The contextual and relational intelligence that experienced treasury professionals bring to their work is not being replicated by current AI—and the gap is not closing as fast as general AI capability progress might suggest.

The Treasury Professional's Comparative Advantage

The clearest thing that emerges from an honest assessment of where AI in treasury is heading is this: the treasury professional's comparative advantage is not threatened by AI in the ways most commonly feared. It is being sharpened.

As AI handles more of the assembly, compilation, and rule application that currently occupies treasury time, the value of what AI cannot do becomes more visible—and more important. AI is not good at knowing why a particular pattern in the data is an artifact of an ERP configuration change rather than a genuine business trend, understanding the relationship dynamics with a key banking partner well enough to know when a proposed instruction will be received as routine and when it will prompt a call, or reading the signals in a CFO's question about liquidity coverage that suggest a strategic conversation is needed before the quarterly review. These capabilities compound with experience. They do not diminish as AI becomes more capable. If anything, they become more valuable precisely because they are the residual that AI cannot replicate.

The organizations that will navigate the next decade of treasury AI most successfully are not those with the most sophisticated models or the most aggressive automation timelines. They are those whose treasury professionals feel genuinely empowered by AI—whose teams understand the systems they work alongside, know when to trust them and when to question them, and are freed by automation to focus on the judgment calls that actually require their expertise.

That outcome does not happen automatically. It requires the deliberate approach to implementation, governance, and team development that this book has described. The technology is a tool. Experienced, thoughtful treasury professionals who understand both the potential and the limits of that tool are the ones who will use it well.

A Note on Pace

The pace of change in AI is genuinely unusual. The gap between what was considered research-stage in 2022 and what is production-deployable in 2026 is wider than most technology transitions of the past decade. This creates a specific challenge for any practitioner writing about the field: the honest caveat that "some of what is written here will turn out to be too cautious" is not rhetorical humility. It is a real possibility.

The response to that uncertainty is not to wait for clarity before engaging. It is to build capabilities that compound—data infrastructure, governance frameworks, team literacy, implementation experience—that will be valuable regardless of exactly how the technology develops. The organizations that are learning deliberately today will be the ones best positioned to take advantage of whatever becomes possible next.

Again, an SAP analogy is relevant here. Over nearly three decades, I watched the platform move through four fundamentally different technical architectures: R/2, R/3, ECC, and S/4HANA. And yet the underlying treasury principles changed far more slowly. Netting, hedging, exposure measurement, in-house bank design, cash concentration, payment controls, counterparty limits—these were refined over time, but the core logic endured. The practitioners who invested in understanding those principles carried that knowledge across every platform transition. The technology changed around them; the principles they had built their expertise on remained useful.

I expect the same will be true here. The specific tools available for AI in treasury are evolving at a pace that makes precise prediction unwise—what is current today may look quite different within a few years. What I am more confident about is that the principles of a well-designed agentic system—clear boundaries, defined handoffs, policy clearly stated, human judgment at the points that matter—are the same principles that have always made complex treasury processes trustworthy and auditable. An investment in those principles will serve you across whatever the technology becomes next.

The profession has navigated every previous wave of technology transformation by combining financial expertise with a willingness to learn what new tools can do. The professionals who built treasury into what it is today brought exactly that combination. There is every reason to think the same approach will work again.

For the Practitioners Reading This

A few thoughts to close with, in the same spirit as everything else in these pages—not as prescriptions, but as observations from someone working through the same challenges.

Start somewhere specific and bounded. The practitioners who make the most progress are not those who attempt to implement a comprehensive AI strategy across the full treasury function at once. They are those who choose one well-defined problem with available data, build something that works, learn from it honestly, and use that foundation to move forward. The framework in Chapter 11 is a structure for exactly that approach.

Invest in data before models. Every AI implementation eventually encounters data quality issues. The organizations that address these proactively—before building models—consistently move faster and achieve more durable results than those who address data readiness before the build begins. The data readiness assessment in Chapter 8 is the right starting point.

Be honest about what you do not know. The organizations that navigate this territory most effectively are those that acknowledge uncertainty openly—with their teams, their CFOs, and their technology partners—rather than projecting confidence they do not yet have. That honesty creates the conditions for genuine learning and course correction when results are mixed.

Bring your team along deliberately. AI implementations that do not have genuine engagement from the treasury professionals who will use the outputs rarely achieve their potential. The override is more important than the model. The practitioner who understands what the system is doing and engages with it intelligently contributes something that no algorithm can replicate. Creating the conditions for that engagement is a leadership responsibility, not a technical one.

Share what you learn. The field of AI in treasury is genuinely young. The practitioners who are building implementations today, discovering what works and what does not, and sharing those findings honestly—through professional networks, conferences, and conversations with peers—are the ones who will shape how this technology develops in the profession. The knowledge that matters most right now is not in research papers or vendor documentation. It is accumulating in the hands of practitioners doing the actual work.

The next few years will likely bring developments that none of us can fully anticipate. Some approaches that appear promising today will not deliver on that promise. Others may exceed what we currently think is possible. The treasury professionals who engage with this technology thoughtfully—with realistic expectations, genuine curiosity, appropriate skepticism, and the patience to build carefully—are the ones most likely to look back and feel they navigated this moment well.

The treasury function has always adapted to new tools, new demands, and new complexities. The professionals who built it into what it is today brought exactly the combination of financial expertise, sound judgment, and practical wisdom that no technology can replace. Those qualities are as important now as they have ever been—perhaps more so, because the decisions about how and where to apply AI in treasury require exactly that kind of experienced, grounded judgment.

The next chapter of this story is yours to write.

Acknowledgements

The practical knowledge in this book—the process challenges, the business realities, the implementation lessons that only reveal themselves when theory meets practice—came from nearly three decades of work alongside treasury professionals, finance leaders, and technology practitioners across some of the world's most forward-looking and dynamic organizations. The treasury professionals and clients who shared their problems, their systems, and their hard-won experience with me over those years are too numerous to mention individually. They know who they are, and this book belongs as much to them as it does to me.

Brad Larson contributed the Foreword to this book, and in doing so brought the perspective of someone who has led treasury at the highest level of global business. He also gave generously of his time in reviewing the manuscript, and his perspective as a senior treasury practitioner shaped the final work in ways that matter. More than that, he brought twenty-five years of friendship, candor, and unwavering encouragement. I am grateful for all of it.

Brian Cody, CEO of Scholastica, reviewed the manuscript with the eye of someone who has built a successful, fast-growing business from the ground up. His perspective—focused on what genuinely matters when managing cash and treasury in a dynamic organization and shaped by the practical realities of entrepreneurial leadership in the digital age—helped ensure that this book speaks to a broader audience: business leaders and treasury professionals alike, in organizations at every stage of their journey.

The proof-of-concept work described in this book was built in close collaboration with **Rohit Todwal** and **Ravi Vyas** and their team at Startbit IT Solutions, Valorean's technical partners, whose work establishing the proof-of-concept architecture and end-to-end connectivity with a test SAP system was an essential step in moving from concept to a working prototype.

My thanks also to **Liz Wheeler** for her careful and thoughtful copy editing, and to **Scott Moden** for his creative cover and book interior design.

To my family— Thank you for the love, affection, and steadfast support throughout what has been a long and fulfilling journey. None of this would mean anything without you.

About the Author

ARJUN KRISHNAN, CTP, is the Founder and Chief Executive Officer of Valorean Technologies Inc., a startup building AI-powered solutions for enterprise treasury.

Before founding Valorean, Arjun spent nearly three decades in enterprise treasury and finance, most recently as a Managing Director at EY, where he led the firm's SAP Treasury Practice in the United States. Over the course of his career, he has advised some of the world's most forward-looking and dynamic global organizations—greenfield implementations, global rollouts, post-merger integrations, and the shift from legacy systems to modern cloud-era platforms. His work has consistently sat at the intersection of deep technical architecture and the practical realities of running a corporate treasury.

Arjun is a Certified Treasury Professional (CTP) and a qualified accountant. He is the author of *Introducing Treasury and Risk Management with SAP S/4HANA* (Rheinwerk Publishing, 2021) and co-author of *SOX Compliance with SAP Treasury and Risk Management* (SAP Press, 2009). He is an occasional speaker and panel member at treasury and financial conferences.

He lives in Raleigh, North Carolina, and can be reached at **arjun@valorean.ai** or through **www.valorean.ai**.

www.ingramcontent.com/pod-product-compliance
Lightning Source LLC
LaVergne TN
LVHW010603110826
845149LV00003B/755

* 9 7 9 8 9 9 6 1 3 5 1 0 3 *